T0274230

Praise for Stuart Sandeman

"The UK's leading breath expert."

—Metro

"[I] was stunned by how well his magic worked."

—The Times

"The most important wellbeing tools are often right under your nose."

—The Telegraph

"Combining ancient wisdom with modern science, he aims to help readers find self-discovery, healing and transformation through guided breathwork exercises."

—The Spectator

Breathe
In
Breathe
Out

Breathe In Breathe Out

Restore Your Health, Reset Your Mind and Find Happiness Through Breathwork

STUART SANDEMAN

HANOVER SQUARE PRESS

ISBN-13: 978-1-335-43068-7

Breathe In, Breathe Out

First published in 2022 in Great Britain by HQ, an imprint of HarperCollins Publishers Ltd.
This edition published in 2022.

Illustrations by Andrew Joyce.

Page design by Nikki Ellis.

Hanover Square Press
22 Adelaide St. West, 41st Floor
Toronto, Ontario M5H 4E3, Canada
HanoverSqPress.com
BookClubbish.com

Printed in U.S.A.

To Tiff. Thanks for guiding me.
And to you, the reader. Thanks for being here.

CONTENTS

INTRODUCTION

How are you breathing right now?

Don't change it in any way.

Just observe and think about it for a minute.

Are you breathing through your nose?

Are you taking shallow breaths?

Does your chest expand more than your belly?

Is there any tension in your body as you breathe?

These may seem like trivial questions, but the answers can tell you a lot. How much energy you have. How much stress you're under. What your overall emotional state is. They even hold the key to healing trauma.

And that's just your breathing over the last 60 seconds.

This is a book about breathing. It's about how the power of breathing can help you take control of your day, let go of your past and make you the best version of yourself. It's about the lessons I've learned from personal experience, ancient wisdom and hard science. And it's about how people from all walks of life have found that something they do all day, every day, can rewire their brains and change the way they think and feel.

In this book you'll find exercises to charge you up, chill you out and improve your performance in any field. You'll learn how to develop better focus, boost your creativity and find your flow. You'll be given tools to help you lessen stress and anxiety, reduce pain and overcome loss. And you'll see how you can use your breathing to release you from the habits, patterns and beliefs that are holding you back.

You'll also find stories: of people who've come from the brink of despair to rediscover happiness, of people who needed something more in their lives—to conquer fears, fulfill visions or meet goals. And you'll hear my story, because I didn't come to this with an open mind. I came as a skeptic. I was doubtful that something seemingly as ordinary as breathing would have the impact it did. Many of the things I've learned in the years since have challenged my preconceptions, forcing me to think a bit harder about how I live. About how we *all* live.

You can learn to breathe your way to better health, better performance and a better life. Through this process I've seen some incredible transformations and been exposed to some fascinating ideas. Some of them are well researched, some anecdotal, some explored but only as yet hypothesized. And although I've come to realize that many cultures have understood the healing power of breathing for thousands of years, my interest is in what can be proved. I've grounded the bulk of my understanding of what has become my life's work in science and practice. This book is designed to reflect that. It's practical and scientific, accessible and fun. Throughout, I will introduce you to a number of breathing exercises that will change the way you think and feel. Some will be quick fixes; others will be daily practices for longer-lasting results.

Whoever you are, this book will help you to thrive in your life physically, mentally and emotionally. It will help you join the dots between how you breathe and how you feel, think, act and perform. It will show you how to reset dysfunctional breathing patterns and give you ways to use breathing to control your emotions, rather than let them control you. It will explore how your breathing changes in response to pain, stress, fatigue, fear, grief, trauma and sadness.

So here's my promise to you: whatever your experience, you can learn how to breathe your way to better health, better performance and a better life. You can learn to develop a closer relationship with yourself, with others and with the world. You can leave behind the things that haven't been working for you, and flourish.

Let's get started.

BREATHE IN.

BREATHE OUT.

You can learn to breathe your way to better health, better performance and a better life.

LET'S BEGIN...

Every author has a hope for how their book might change their readers' lives. Often it's rooted in their own experience—some event that transformed them in some way. That was the case with me. I went through something that changed my life and made me look for answers.

HOW I STARTED BREATHING AGAIN

I squeezed her hand three times. It was our code. *I'm here. I care. I love you.* I had to be strong for her. But as I stared across the desk at the doctor reading his notes, I was barely breathing. My knee bounced restlessly.

It was only a couple of months earlier that my girlfriend, Tiff, had found a pea-sized lump on her chest. Until that day we'd been having the time of our lives. She was 30, and a fun-loving fashion buyer. I was 31, a DJ without a care in the world. But then cancer joined our party and dragged the needle violently across the record of our lives. Now we were here, sitting in silence in the oncology unit of the UCLA Medical Center in Los Angeles, waiting for the specialist to give us the news.

It may only have been a couple of seconds before the doctor spoke, but it felt like hours. He took a deep breath, preparing us for what he was about to say.

"The scans confirm your cancer has metastasized. We've found tumors in your liver, spleen and brain."

Western culture does not encourage us to face death when we're still very much alive. So it's a hard thing to process when you become aware that someone you love might die. I did what I'd always done: hid signs of weakness, buried my emotions, coped the only way I knew how. *Be strong. Be tough.* I turned to Tiff. There were tears standing in her eyes.

"We'll beat it," I said.

Being strong came naturally to me. I'd grown up in Scotland with a teddy bear called Tough Ted. I had *Rocky* posters on my wall. I'd trained in judo since I was four and was a black belt by 16; I was Scottish champion for years. I lived my life this way. Cancer was just another opponent.

As for Tiff, she was from Taiwan, but grew up in New York. She was smart, incredibly well-read, streetwise. She knew how to hustle. Together, we were determined to prove the doctors wrong.

In the year that followed that conversation at the UCLA Medical Center, we did everything we possibly could. We traveled from LA to New York, from London to Taipei, searching for a cure. We saw experts, consultants and doctors. We even met healers, shamans and monks. We promised each other that we'd be open to anything, that we'd try everything. Yet nothing seemed to work. Her health started to fail.

And then, six months later, something miraculous happened. It looked as if the cocktail of chemo and surgery, juicing and meditation, was paying off. Tiff started to show signs of recovering. The doctors said they'd never seen anything like it. She seemed to be coming back from the very brink of death. Her brain tumors had gone, her seizures had stopped and her smile had returned. It was as if the thing no doctor had said was possible—that Tiff would actually *get better*—was about to happen.

Encouraged by this, I left her bedside. I needed to fix the leaking roof of my apartment; I had been putting it off for months, and it wouldn't take long. But as I was on my way back, flowers in hand, I noticed four missed calls and a text from Tiff's mum. My heart fell. Somehow, I knew what it would say before I opened it.

"Get back ASAP. Tiff's heart has stopped."

And then:

"The doctors can't do anything."

It was February 14, 2016. She took her last breath on Valentine's Day.

A CHANCE ENCOUNTER

When Tiff passed away, I couldn't think. I couldn't feel. I busied myself with all the practicalities of death: supporting her mum and arranging the funeral. When it was all over, I shut down. I didn't know how to express my emotions. I didn't know how to cope with loss. I bounced between outbursts of anger and complete withdrawal. I thought that pushing everything and everyone away was a way of dealing with my grief, but it wasn't. I couldn't engage with the world or find my place in it without Tiff. The image of strength I'd always projected was beginning to show cracks. I needed something else.

I'd always thought of myself as a logical thinker, and I considered this a good thing. I came from the world of performance and science. As well as training in judo, I studied mathematics at university, and went on to work in finance. Even when I made a dramatic career change, swapping the chaos of the trading floor for the nomadic lifestyle of a DJ, I did it with my eyes open: gathering information, assessing risk, making the change and then observing the results. That's how I did things. What Tiff's death showed me, in the most tragic way I could ever imagine, was that there were shortcomings to this approach.

It was a long eight weeks after Tiff's death that I found myself, by chance, in a breathing workshop with my mum. It was a gift I'd bought her for Mother's Day. We took our shoes off and entered a room bathed in natural light, with high ceilings and colorful embroideries on the walls. Ambient music filled the space; the plants and statuettes seemed to be singing. I could smell the burned-pine scent of palo santo, the "holy wood" incense shamans had used in Tiff's healing sessions.

"Hello," said a smiling woman dressed in white. "You must be Stuart. Come, take a seat in the sharing circle."

My gaze fell on the rest of the group. My heart sank. *Jeez,* I thought. *I hate this sort of thing.*

Can you blame me? Over the past year I'd seen enough medicine men and healers to last a lifetime, and all of them had claimed to have miraculous cures for whatever ails you. My tolerance for anything that seemed even remotely "spiritual" had worn thin—and that was *before* the "heart-shaped stone" was pressed into my palms and I was asked to "share my intention."

"I'm Stuart," I said, sheepishly. "I recently lost my girlfriend to cancer. So I intend to feel, um, a little lighter, I guess."

Despite my cynicism, I was grateful for the looks of love and support I received after saying this, even if I felt pretty awkward at the same time. But if everything that had happened so far during the workshop wasn't enough to take me out of my comfort zone, what came next certainly made sure of it. I was shown a method of breathing that involved lying on my back, and alternating between heavy breathing and what looked like having a tantrum. *If only Tiff could see me now*, I thought.

On went some New Age trance music and everyone in the room began to huff and puff. I opened one eye to check the whole thing wasn't some elaborate prank. But my mum seemed to be getting into it, and the workshop was my gift to her. I wanted to be there for her, just as she'd been there for me. All I had to do was play along. In for a penny, in for a pound.

After a couple of rounds of breathing and shaking and shouting, something pretty bizarre happened. I could feel electricity surging through my entire body, the kind of vibrations you feel when standing in front of a giant festival speaker. Lights danced and flashed behind my eyes. A giant wave of emotion roared up inside me. And then, for the first time in as long as I could remember, I cried. I cried and cried and cried. And not only did I feel the weight of grief being pulled off me, but I felt as if a lifetime of tension I'd unknowingly been carrying around was dissolving into the atmosphere. I felt a strong presence surrounding me and had the very distinct feeling that Tiff was there, holding my hand. I still get goose bumps when I think about it. It was weird. It was powerful. It was life-changing.

Even in my grief, I knew that my experience wasn't logical. Nothing about it made sense. My rational side could only see two possibilities: I'd either completely lost the plot or someone had slipped a hallucinogen into my drink.

I asked one of the facilitators about what had happened. Was my experience normal? What had I just gone through? Her amethyst earrings glinted in the light. She smiled.

"You connected with spirit" was all she said.

That response might have been enough for some people, but it wasn't enough for me. I needed to know what had happened and in a way my brain could understand.

I had questions that needed answering. It was time to get to work.

THE WORLD OF BREATHING

I never could have imagined that something as simple as breathing could heal my grief and transform my life. I mean, really? That thing we do all day? How could doing that differently change a thing? I'd never thought of breathing as a tool or practice. I'd always been too *busy* to breathe. And anyway, if someone had told me to *just breathe* through Tiff's cancer, they'd have got either a mouthful or a serious eye roll from me.

And yet in the year that followed that Mother's Day session, a regular breathing practice didn't just release me from the pain and uncertainty of grief. My energy also increased, my mind cleared, my fitness levels went through the roof. My nightmares stopped and my sleep became deeper. Even the voice in my head began to sound a little kinder. I felt able to move forward. I felt hopeful for life again.

The more I learned, practiced and observed, the more I became convinced. I hadn't lost the plot. No one had spiked my water bottle. This release that I'd experienced, this ability to express and understand the feelings I had about Tiff's death and myself, this ability to connect to a powerful state, deeper than any meditation I'd come across, was down to the power of breathing. If it had helped me, then it could help others too.

I threw myself deep into the world of breathing. I studied a number of breathing modalities—and yes, there are a few: some very practical and scientific, others more spiritual and mystical. I read research journals. I met breathing experts. I hung out with consultants, yogis, healers and gurus. While the world was asleep, I was up watching documentaries, buried in books or deep in practice. I was obsessed.

For a long time, nothing else mattered. All the energy, determination and enthusiasm that had driven my years of judo and my professional career went into this new mission. I felt like Tiff was at my side, urging me on, and I wanted to make her proud. I charted the straightest course I could to help people become happier and healthier through breathing. Within 12 months, I'd set up my own small private practice with the goal of introducing more people to its life-changing power.

Client by client, people underwent miraculous transformations. I watched stressed-out city workers find calm. I saw painfully shy children grow confident. I observed people in the grip of depression become happy and optimistic for the future. I even helped chronic insomniacs sleep through the night. People young and old, skeptical and open-minded, shed the weight of negative emotions they'd been carrying around with them for years.

Despite this, I was under no illusions that the things I was seeing in my clients and had experienced myself were understudied. I knew that many cultures and traditions had a long history of using breathing to help people endure the slings and arrows of fortune. Chinese *qi*, Sanskrit *prana*, Egyptian *ka*, Hebrew *nefesh* and *ruah*, Greek *psuchē* and *pneuma*, Latin *anima* and *spiritus*, Polynesian *mana*, Iroquoian *orenda*… Even in the Bible, God breathes life into Adam. They all highlight the importance of breathing for the body and mind, and its connection to something deeper. But I was not satisfied. I needed to build on the knowledge I was accumulating. If I wanted to help as many people as possible, I had to start filling in those spaces in our collective knowledge and gathering data in whatever way I could. Clearly something powerful was happening to people during these breathing sessions, just as something powerful had happened to me. People were experiencing changes in their well-being, releasing deep trauma, discovering the ability to move forward and gaining new insights about their lives. But pinning down exactly *what* was happening in the body and mind was still unclear. My search for answers wasn't over.

I was encouraged by a world-renowned scientist, and now close friend, Dr. Norman Rosenthal, to log my client experiences and try to make more sense of what was happening in the sessions. The existing information out there was patchy. We wanted to start to fill in the gaps. And though this isn't to say that modern science doesn't understand the healing potential of breathing, there's a lot we're still figuring out. It's been fascinating to see teachings from old traditions, philosophies that once seemed mystical, start to make scientific sense. You'll be meeting Norm and reading about our findings later in the book.

What I've come to understand is that our breath is the bridge between our physical, mental and emotional states. It's a powerful tool to improve our health, heal us from negative events in our past and even access higher states of awareness. It's the key to the door that connects the conscious and unconscious mind. If we can control one, we can control the other.

And breathwork isn't exclusive to the spiritual elite or to the sharing circles. You don't need a guru, master or sensei.

Breathing is yours to own, and when you know how to use it, you can do it anywhere, from a mountaintop monastery to the train you take to work. It doesn't matter whether you use it to relax, to achieve a goal or to transform yourself. You're in control.

Our breath is the bridge between our physical, mental and emotional states.

THE POWER OF INTENTION

There's something you have to do before we go any further. And it has to do with intention.

As I said, I was pretty cynical when I was asked to share my intention in that Mother's Day breathing session. Maybe you are too. But I've since learned that when it comes to making transformative change in your life, you have to start with an intention. It was only when I committed—when I made the intention to "feel lighter"—that things really started to happen. I want this for you as well. A positive intention leads to a positive attitude. A positive attitude paves the road ahead.

Setting intentions allows you to focus your physical, mental and emotional energy. I like to think of it as punching your destination into Google Maps or aiming your sight on a goal or target. If you don't know where you're going, you might just end up driving round in circles. Intentions have the power to change your body and mindset in a positive way. They're the seed of what you aim to create. They help you prepare to be the change you want to see in the world.

With intention you make a deal with yourself to overcome your resistances and commit to change. This isn't always easy. There may be times in this process when you feel unsettled or out of your comfort zone. There may be times when you need to stop reading and simply process what you've learned. There may be times when you need to reread sections. All of this is OK. It's called "breathWORK" for a reason! If you want results, you've got to put the work in.

But this is your moment, your chance to step up. This is your opportunity to learn more about yourself and reach for a physical, mental and emotional level that you may never have thought possible. This is your moon shot. I'll be with you every step of the way, but it's up to you. You can put the book down and walk away now, or you can say, "Yes, Stuart, I'm all in!" and keep reading.

EXERCISE 1
SET YOUR INTENTIONS

Go into as much detail as you like with setting your intentions. They're unique to you. Think about your life, your health, your habits, your work, your relationships, the world around you. Think about your thoughts and your feelings. Think about who you want to be.

I've added some questions to help refine and define them. The answers to these might not come to mind straightaway. But give them your best shot, and keep returning to this exercise as much as you can. You'll find as you get moving through the book that the answers may change or start to crystallize in your mind.

- How do you want to feel when you wake up in the morning? *e.g., well rested, energized and excited for the day ahead.*
- How do you want to think and feel throughout your day? *e.g., focused, optimistic and calm.*
- How do you want to feel when you go to sleep at night? *e.g., relaxed, proud and grateful about what I've done that day.*
- What would you like to let go of and release? And how will you feel when you let go? *e.g., I'd like to let go of my poor breathing habits, my overthinking and worry. Then I'll feel more alive and calm.*
- What are the challenges in your life? *e.g., recurring lower-back pain, turbulent finances and a challenging relationship.*

- What would you like to come into your life? *e.g., more joy, peace and confidence.*
- What do you want to be more of to your family/partner/friends/co-workers/community? *e.g., I'd like to stop people-pleasing and find the confidence to voice my opinion. I'd like to trust people more easily, be kinder and support those around me.*
- What do you need to feel more like yourself? *e.g., more me time doing what I love.*
- What do you want for the world around you? *e.g., I want the world to be peaceful and balanced.*

Your final stage is to affirm your intentions in the here and now. If you're always "intending" or "wanting," then you're always looking ahead to something in the future. It's like dangling the carrot in front of you—it will create forward motion, though you may never actually get the carrot. But if you can transform your intentions into statements that are true now, it will help put the carrot in your hand, creating more energy and focus in the present.

Using your answers to the intention-setting questions, create three statements (or affirmations), making sure that they're all in the present tense and positive (I am, I choose, I believe, and so on).

For example:

- *I am grateful and optimistic.*
- *I choose to be confident.*
- *I take time to do the things I love.*

1

BREATHING, THINKING AND FEELING

BREATHING IS ENERGY

I'd like you to stop breathing right now.

That's right. Hold your breath.

Keep reading until I tell you to breathe again.

You won't be surprised to learn that breathing is fundamental to your survival. You can go days without water. You can go weeks without food. But go about three minutes without breathing, and you're in trouble. Even now, during this relatively short break from breathing, you're probably feeling a little uncomfortable. So, no more torture.

Breathe in.

Your whole life is a dance of breath. It's the first thing your mother listens out for when you're born. It's the last thing your loved ones see you do when you die. Right under your nose, some 20,000 times a day, an orchestra of bodily systems play a symphony as your heart and lungs duet, bringing energy to your body and life to your cells.

Wait, *energy*? I was always taught that *food* is fuel. You probably were as well. You probably think that on your average day, your energy comes from the food you eat. But that's only partly right. The most common and effective way for the food you eat to be broken down into a usable energy source requires oxygen, which requires breathing. So, looked at this way, two-thirds of all your energy actually comes from the air you breathe. And as most of your cellular waste is carbon dioxide, this means that about 70 percent of your body's waste is expelled through your lungs when you breathe out—the rest is removed through your skin (sweat), kidneys (urine) and intestines (stool).

If we strip things right back to basics, that's what breathing is about: energy in, from the air around you; waste out. OK, it's a *bit* more complicated than that. But it's a good place for us to start.

OXYGEN + GLUCOSE = ENERGY

Oxygen in the air is transported to your cells, where it combines with glucose to produce adenosine triphosphate (ATP). It's an energy source that enables your cells to perform their many functions. With it, your brain can send electrical impulses, your heart can beat, your eyes can see, your muscles can contract, you can move and grow. In this process—which is called *aerobic respiration*—carbon dioxide, water and heat are also produced and removed from the body when you breathe out.

WHAT YOU'RE DOING = HOW YOU'RE BREATHING

If breathing gives you energy, then *how* you breathe depends on how much energy you need. We see this in action every day. Let's say you turn the corner after a long day and see the last bus that can take you home about to pull away from the bus stop. You set off to try to catch it. What happens to your breathing? Well, the muscles in your legs need more energy, so your breathing speeds up, and your heart beats faster both to pump more oxygen to those hardworking cells of yours and remove the excess carbon dioxide created in the process. You'll probably need to catch your breath afterward to recover, because you'll still be producing carbon dioxide. And when, later in the day, you get into your bed, your breathing will slow down to calm your mind and let the tension leave your body so it's easier for you to sleep. So your breathing is influenced by what you're doing. That's the first thing.

This change in energy through breath doesn't only happen in situations where we consciously decide that we need to move more or move less. Your brain is always scanning your environment for signs of safety and danger. It's trying to keep you alive. It reads thousands of social and environmental cues in your surroundings and automatically chooses a type of breathing—fast or slow, deep or shallow—that will give your body the right energy it needs to respond to your environment without having to think about it. Your brain also monitors the signal it receives back from your breathing pattern to trigger a physical and emotional response: alertness, distress, relaxation, attention, excitement or anxiety. Even when you interact with others, you pick up on facial expressions, tone of voice, bodily movement and much else. This changes your breathing too.

All of this happens because there's a link between your breathing and something called your autonomic nervous system (ANS). This connection enables your breathing to become short and shallow in stressful situations, and your heart to pound and pump oxygen to your muscles to get you ready for action. It's one of those alarm bells our species has had for millions of years to get us out of danger. The ANS also allows you to take long, drawn-out breaths to rest, recover and repair your cells when you're safe.

Sounds technical? Well, don't freak out. Instead, let's geek out! Armed with this knowledge, you'll soon understand how effectively your breathing can change the way you think and feel.

First, though, let's take a look at the brain.

BREATHING IS THE BRIDGE

It doesn't matter if you're a perennial perfectionist, an outgoing optimist, an introspective introvert or something else altogether—your personality and behaviors are derived from the different levels within your mind: conscious, subconscious and unconscious.

The **conscious mind** contains the thoughts, memories, feelings and wishes that you're aware of at any given moment, including the things that you know about yourself and your surroundings. It's the awareness that you're experiencing right now as you read this. This is the aspect of our mind that we can think and talk about with others.

The **subconscious mind** is just below the surface of our awareness. It contains anything that could potentially be brought into the conscious mind at any given moment. For example, if I asked what you had for breakfast or your pet's name or what your house number is, you'd be able to recall it. Your subconscious stores your memories and remembered experiences. When you develop a habit or practice something repeatedly, it gets stored in your subconscious and becomes part of your behavior.

The **unconscious mind** contains thoughts, repressed feelings, hidden memories and habits outside of your conscious awareness. It's the source of your primitive instincts, urges, desires and motivations. It's where you store childhood memories, emotions and even negative experiences that are too painful, embarrassing, shameful or distressing to face. Your subconscious mind buries these deep within the unconscious mind as a way to protect you. If your brain is a house, your unconscious mind is the locked cellar under the stairs. It stores the core beliefs, fears and insecurities that drive your behaviors today.

What does this have to do with breathing? Well, as I mentioned, breathing is one of the very few vital bodily functions that's controlled by both the conscious and unconscious parts of your mind. You can direct its flow—speed it up, slow it down or stop it altogether using your *conscious* mind—or you can leave it to run all by itself and let your *unconscious* mind take over (which is pretty handy, because on your busier days you might otherwise forget to do it). While you will most likely relate to the functions of the conscious mind, your unconscious mind is, as the name suggests, the part of your brain that you're very unlikely to be aware of, despite it affecting your behaviors, thoughts, feelings and breathing every day.

A TALE OF TWO SIDES

Your autonomic nervous system (ANS) is controlled by your unconscious mind. It is the way in which you breathe that influences your ANS and causes you to move between two states: active and at rest. These two states are determined by which division of your ANS is more dominant. For example, when a threat is detected, you trigger a division of the ANS called the *sympathetic nervous system*. Think "S" for "stress," the state commonly described as "fight or flight." This stressed-out state triggers and is triggered by fast-short-shallow breathing. It also happens when you engage in physical exercise (like running for the bus). On the flip side, you have the division of the ANS known as the *parasympathetic nervous system*. Think "P" for "pause," the state commonly described as "rest, digest and repair." This chilled-out state triggers—and is triggered by—slow, gentle, deep breathing. It happens when you're relaxing, recovering and sleeping.

Once you understand and recognize these divisions and the corresponding physical, mental and emotional states they produce, you can use your breath to bounce between them or balance them out. It's important to recognize that neither state is better than the other. You need to use them both, depending on the situations you find yourself in. However, you may find yourself stuck in one state or too often using one over the other. Maybe you feel like you're always "on" and need to hit the "off" button, or that you've lost your get-up-and-go and need a boost of energy. Or perhaps you simply need to find some balance and learn to hit cruise control.

EVEN THE BUDDHA FELT STRESSED

Stress is unavoidable. Even the Buddha felt it. In a sympathetic state, your brain's fear center, the amygdala, takes charge and signals to your adrenal glands to release stress hormones into your bloodstream to shift your body into gear. This drives your heart to pump faster and harder, raises your blood pressure and engages your muscles. Your pupils dilate and your breathing tubes open to get more air in. Your breathing becomes fast, short and shallow as you redirect oxygen-rich blood to areas of the body that need it during intense physical demand. You break into a sweat. This response also makes you act in a more reactive way. Your senses change. You view things, including people, as more of a threat. Even your sense of hearing changes, becoming more acute.

When your sympathetic nervous system is dominant, or you're stuck in it, anything that's not required in an emergency goes on "airplane" mode: your digestion, sexual organs, even your higher executive brain functions—reason, memory and creativity—shut down. All your energy goes into keeping you alive.

This kind of response was once crucial to our survival as a species. If you stepped inside a time machine and traveled back 40,000 years to when your ancestors were nomadic hunters and foragers, you'd find that the feelings they experienced were just like yours. Their lives were tougher than yours—no running water, no food delivery, no Netflix and chill—but in many ways, they were more straightforward. Their days were mostly about finding food and keeping safe.

Keeping safe often involved avoiding predators. Our ancestors' breathing would have played a large part in this. If they came across a grizzly bear while picking berries, their senses—what they could see, hear and smell—would channel information to their brain. Their brain would, in a split second, analyze and register the threat, then trigger the sympathetic system: their respiratory control center would increase their breathing rate and facilitate an oomph in energy, their bodies would be flooded with stress hormones, their hearts would pound and their brains would go on high alert. Their digestive and reproductive organs would shut down (as would anything else not needed in a bear emergency), and oxygenated blood would flow to their muscles so that they could either run to safety or brace themselves for a fight, which—let's face it—they'd probably lose. If you're always being bombarded with notifications or struggling to work through your to-do list, you can get stuck in that stressed-out, sympathetic state all the time, and it's like you're running from grizzly bears all day long.

TAKE A CHILL PILL

The parasympathetic state, on the other hand, is the one that everyone wants to be in—probably because we're all suffering from too much stress in our fast-paced world. Who doesn't want to be cool, calm and collected?

This state is commonly referred to as your "rest, digest and repair" response. Your breathing rate slows, enabling your brain and body to conserve energy. Your heart rate slows down, your blood pressure drops, your muscles relax, digestion begins, and your cells repair and regenerate. This mode helps you digest food, sleep deeply, have sex, feel safe and relaxed. When the parasympathetic drive dominates, your brain's frontal lobe is also activated, which plays an important role in higher cognitive functions. It's critical to planning, decision-making, creativity, reasoning and judgment. In other words, in a parasympathetic state you're calm enough to be able to consider your response to external stimuli, instead of resorting to knee-jerk reactions.

If you spend too much time in this state, however, you can lose your motivation and nothing will get done. It's often caused by a traumatic experience or a long and difficult period, such as when you're struggling to find a job. I've had clients suffering from depression, disconnection and lethargy who became stuck in a parasympathetic state. But again, it works both ways—getting stuck in this state can also lead to depression and other withdrawn emotions. Let me explain this in more detail through a third response.

PLAYING DEAD

There's a third major response that you might have heard of, or even experienced: freeze. It's another way that our ancestors may have responded when faced with a grizzly bear; they'd freeze, become immobilized and play dead. Sometimes, that really would be the safest option. We still have this response today. Technically it's a stress response, but it belongs to the parasympathetic arm of the ANS. You freeze when your unconscious brain believes that neither fight nor flight is available to you and you're totally overwhelmed or trapped. This isn't a conscious decision. The primitive brain takes over, and makes you immobile and numb. The intense rush of stress causes the parasympathetic system to kick back in to immobilize you: your heart rate slows, and breathing shortens or halts altogether. The sudden braking can even make you faint. It's your brain's last hope that whatever the threat is—like that good old grizzly bear—it will lose interest in you and wander off, or, if you do end up in its mouth, you won't feel anything.

But what role does this reaction play in modern life? It's been suggested that freezing is a form of psychological protection that can block your ability to consciously experience something overwhelming or traumatic. You completely shut down and withdraw. That way, you might not be able to prevent something from happening, but you stand a chance of saving yourself from feeling pain by locking it deep within your unconscious mind. Sometimes, those that experience a traumatic childhood live in this shut-down state. As a result, as an adult they can barely recall this period of their life.

When someone is depressed, they're often stuck in this state, a permanent condition of parasympathetic freeze. This can also happen to a lesser extent in social situations, where some people find that they move from being anxious into a state of complete withdrawal. I used to experience a debilitating version of this freeze response when I had to do any form of public speaking at school or uni. I even turned down a chance to do a wedding speech for a good mate because of it.

Both the sympathetic and parasympathetic divisions of your nervous system are always active, but one is often more dominant. It's the natural interplay between both that enables your breathing and your cardiac system to respond quickly to different situations. It's a bit like a tug-of-war. Your sympathetic state is always tugging you to high alert, looking to make your breathing quicken and your heart rate increase to prime you for action. Think of it as the overprotective, hypervigilant friend who loves you so much they want to keep you safe at all times—"Don't do that, watch out for that, let's get out of here, that's not safe."

Your parasympathetic state is the opposite. It's looking to slow down your breathing and lessen your heart rate. It's the friend who loves you so much and just wants you to be calm, to relax and have time to digest your food—"Just chillll, relaaaaax, have a seat, take a break, digest your dinner and get some rest." You need both of these friends in your life at different moments.

While you're very unlikely to run into a bear today, these instinctive responses continue to affect you on a daily basis.

Let me give you an example of how these might play out in your day. You've had a manic morning, so you drop everything and pop out for lunch. There isn't a cloud in the sky—for once—and it's the perfect day for a relaxing stroll. You buy a sandwich from somewhere on the high street and head toward a nearby park. And though the sandwich is underwhelming (and overpriced), your parasympathetic state dominates. You're relaxed.

You reach the road opposite the park, look left and right, and prepare to cross. But just as you step out, you hear a loud "ding, ding, ding!" You whip your head round and see a flash of red. A cyclist has appeared out of nowhere. Your sympathetic state kicks in and stress hormones flood your body.

Without thinking, you take a large gasp of air through your mouth—an early trigger for your sympathetic system. Your digestion stops and your heart pounds, sending blood rich in oxygen to your leg muscles. Your cells turn this oxygen into energy, your leg muscles contract and you leap back onto the pavement, to safety. Here you breathe a huge sigh of relief, flicking you back to your parasympathetic system. Your heart rate decreases, you begin to calm down and return to digesting your (underwhelming) lunch.

Even when you're not in a physically dangerous situation (like crossing a busy road), these responses still affect you on a daily basis. This is because your unconscious mind cannot distinguish between a stressful experience happening around you and a stressful experience that's fabricated through your thoughts alone. Both experiences trigger the same response. This means that if you're continually ruminating on the past, rehashing old mistakes or worrying about the future, you can become stuck in a sympathetic stress response with stressful breathing. You can even get stuck in this stressful breathing pattern, which is like constantly ringing the alarm bell to your brain. Equally, if you're feeling withdrawn, unwilling to get out of bed until midday and are completely unmotivated, you can get stuck in the parasympathetic state.

THE NEGATIVE-THINKING LOOP

In 2005 the US National Science Foundation published an article summarizing research on the number of thoughts someone has per day. It was found that of the 12,000 to 60,000 daily thoughts the average person has, 80 percent are negative and around 95 percent are repeated thoughts. This unconscious negative bias, or a tendency to sway toward negative thoughts, isn't some human design flaw. Quite the opposite—it's another survival tool, a hardwired instinct. Have you ever woken up in the middle of the night, thought someone was in your room, freaked out, switched the light on and found that it was just a coat hanging on the back of the door? We're all on high alert, and as a species, we have been for thousands of years, assessing for risks: those tigers and bears that don't usually threaten us anymore. Unconscious negative bias is how we're programmed. It means that the majority of us are stuck in a loop of negative thinking that triggers stressful breathing patterns.

BREATHING, THINKING AND FEELING

Studying the ANS in this way reveals the most fundamental principle of breathwork: that not only does the way you breathe affect the way you think and feel, but also that *the way you think and feel affects the way you breathe.* It's a feedback loop. I'll walk you through it. Thinking happens in the mind. Feelings happen in the body triggered by your breath. Your thoughts and feelings change the way you breathe, and your breathing changes the way you feel, which changes the way you think.

When both your thinking and your feelings match, it creates your state of being. So if you're thinking anxious thoughts and feeling anxious, you're in the sympathetic mode. Your heart is racing, you're taking short, shallow anxious breaths, and your state is "I'm anxious."

In most cases you can break this "state of being" loop in one of two ways. You can either think differently, accepting or replacing the anxious thoughts with more positive ones, which can be difficult when you're anxious. Or you can breathe differently to change how you feel in your body, so it no longer matches your thoughts and your "state of being" loop is broken. Your brain says, "Hold on, I'm *not* anxious. My heart isn't racing, my breathing is relaxed." This "state of being" loop applies to any state, positive or negative. For example, breathing calmly will create a feeling of calm, and calm thoughts will follow.

Throughout this book, we will work on doing both thinking and feeling differently. Because here's the thing: if you're stuck in a loop, you're stuck in a state of being. You may even start to identify with that state of being as part of your personality. This is because a state of being that lasts a week becomes a mood. A mood that lasts for months becomes a temperament. A temperament that lasts for years becomes a personality trait.[1] This means that some personality traits can be traced all the way back to the way you've been breathing. In other words, a part of your character that feels permanent and deeply personal could have begun with just a single breath.

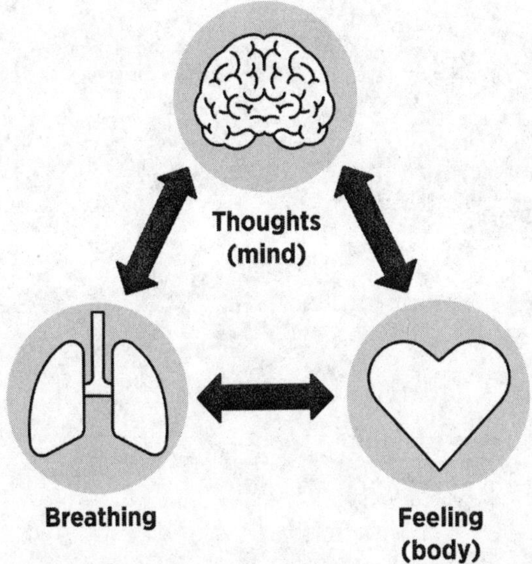

*Please note it's important to recognize that there will be cases where very strong negative emotion is caused by biological factors or imbalances. If you're experiencing serious distress, please do seek professional support.

All this shows just how important it is to pay attention to your breathing. Because by controlling your breathing, you can start to reconnect with your unconscious mind, take control of your thoughts and feelings, and make positive changes in your life.

But, as your unconscious mind has automatic control over your breathing, we need to explore what has been programmed into it.

When both your thinking and your feelings match, it creates your state of being.

YOUR BAG OF BRICKS

Ever seen a baby breathe? Babies are the breathing gurus. Watch a baby breathe and you'll see an example of the kind of perfect breathing that yogis, singers, actors and other performers spend years training to achieve. Unless a baby is crying, they'll only breathe through their nose. With each breath in, their lower torso expands like a balloon. Their wee belly puffs up first, then their little ribs and then their chest. Their exhalation is relaxed. It's efficient. It's effortless.

From that first moment, it seems as if your breath anchors life in your body and you set off on your way. But as you'll soon learn, most people are dysfunctional breathers, whether they know it or not. So what changes in the years that follow? If most of us are born as perfect breathers, what happens between infancy and adulthood that turns us from masters to amateurs? Well, life happens. Throughout our lives, we all have experiences that affect the way we breathe. It's not just the stressors and negative thinking. The way we breathe could be impacted by other factors—posture, injury, illness, to name a few.

Very rarely do we think about the impact that our posture has on our breathing. Similarly, we rarely think of the impact a stressful job has on the way we breathe, or—the reverse—the effect our breathing has on our stress levels. And of course, most of us *never* give a thought to how our breathing is affected by the different experiences we've had over the course of our life, whether that's being shouted at for failing to do our homework as a child, falling off our bike, being blamed for something that wasn't our fault or being involved in something more traumatic, such as a breakup, divorce or bereavement. All of your past experiences, however important or trivial they may seem, have an effect on your breathing today.

Let me give you an example.

Meet my sisters, Jane and Anna. When Jane sees a dog, she registers it as a threat. She sees the dog looking at her, baring its big, sharp teeth, and she feels that it's about to attack her. Her sympathetic system is activated. Stress rushes through her body, her breathing rate increases rapidly and her heart sends oxygenated blood to her legs. She makes a dash for the door.

Now consider Anna. When she sees the same dog, her face lights up. She sees it smiling and smiles back. She remembers the childhood joy of playing with her friend's family dog, Milo, excitement charges her body, her breathing flows openly. She runs off to go and give the dog a scratch behind the ears.

What's happened here? Why do Jane and Anna behave so differently in an identical situation? Clearly, Jane is scared of dogs. When she was much younger the neighbor's yappy terrier nipped her on the hand when she went to stroke it. As for Anna, she has happy memories of rolling around in the garden with Milo. She sees all dogs as loyal and loving.

Your life experiences shape your perceptions and consequently how you breathe and operate in your world. Even if Jane and Anna were identical twins, their life experiences would make them unique and cause them to form different beliefs. And this, inevitably, would inform how they breathe.

So as life begins, it's as if you start off with an empty bag on your back. As you go through life, your experiences are like items thrown into the bag—pebbles, boulders and bricks. Sometimes, other people start slam-dunking their bricks into your bag too.

We all have this bag. Even the most experienced Zen monks have this bag. (Of course, their bags are lightweight and neatly packaged.) For most of us, the bag gets heavier and heavier over time, so heavy we can't even see where we're going. Maybe it even overflows. Life becomes a struggle and we can't move forward anymore.

The pebbles, boulders and bricks get heaped into that bag in the form of physical tension. When you're overwhelmed, you can pretend the bag doesn't exist or that it isn't as heavy as it really is. You can pretend, in other words, that by suppressing or repressing your emotions, they have gone away. But this is a safety mechanism, protecting you from an outburst or from feeling emotional pain. This process happens unconsciously, and it takes over your breathing.

You've already learned that breathing is about energy, and how it's the bridge linking the conscious and unconscious parts of the mind. Left to run on its own, it triggers one of the two divisions of the autonomic nervous system—sympathetic or parasympathetic. These affect how you think and feel: "on" or "off," stressed or chilled. But by controlling your breathing, you can choose which state of the ANS to be in, and so change your state of being.

What this all means is that by understanding how you are breathing and learning the basics of breathwork, you can put that bag of bricks down for a moment. You can learn to respond instead of react. You can take charge of how you feel in any situation. The exercises in Part 1: *Fix*, will help you make some quick changes to turn your stress into calm, your overwhelm into balance, even manage your pain.

Of course, life will still happen, and you'll have to pick the bag up again. But with more practice using the exercises in Part 2: *Deeper Work*, you'll learn how to empty the bag completely of all the past experiences affecting you today and wipe the slate clean. You'll be able to rewire who you are, go beyond dealing with the daily stresses of life and get to the root cause of whatever you're struggling with today. And when you've done this—when that bag is empty—you will feel lighter, sleep better, have more energy and experience less stress. You'll be the best version of yourself, unburdened by negative past experiences, reactive patterns and limiting beliefs.

Once you've mastered how to empty your bag, you can start looking at how to enhance your breathing. In Part 3: *Optimize*, I will show you how to harness your breath to improve your performance and focus in all areas of life, so that you can achieve more, and be more present and connected with the world around you.

To bring this all together, you now need to start paying close attention to how you're breathing. This can tell you a lot about how you're thinking and feeling, and that opens the door to change.

PART 1: FIX

2
WHAT YOUR BREATHING SAYS ABOUT YOU

HOW ARE YOU BREATHING?

As a species, we seem to have lost our ability to breathe properly. You may assume that there's nothing wrong with your everyday breathing. But most of us have picked up bad breathing habits, and we all have past experiences stored in our unconscious mind that constrict our breathing. It's almost impossible for you to feel good, think straight, sleep well, overcome illness or benefit from exercise or health regimes if your breathing is out of whack. If you're not breathing properly, your cells, brain and organs do not get the fuel they need for optimal health. Your body must work overtime to release toxins. Bad breathing makes us tired, stressed, anxious, and impacts our physical, mental and emotional health.

The good news is that it can be fixed.

It all starts with awareness.

To be aware of your breath is to observe your unconscious breathing pattern. Awareness of your breathing involves peering through the keyhole of your unconscious mind, giving you an insight into how your brain and body are reacting to their perceived reality. By mastering breath awareness and applying it throughout your day, you will know what exercises you need to do to change your breathing and alter how you feel.

Breath awareness can be done anywhere and in any situation. So let's give it a go right now. The most important thing is not to try to change your breathing while you're doing this exercise. Be a neutral observer. Be a breath detective. Try to pay attention to every little detail as I guide you.

EXERCISE 2
BREATH AWARENESS

- Pause. Take a moment.
- What's your posture like? Are you hunched over or tense? Is your chest open?
- What's your clothing like? Is it loose or restrictive?
- Now feel the air around you. What's the temperature of the air? What's the texture of the air?
- Follow the air into your body. Are you breathing through your nose or mouth?
- Don't try to change anything just yet. I want you to be the observer of your own breath.
- How does it feel just to be and to breathe?
- What muscles are working to bring air into your body?
- What parts of your body are moving as you breathe in? What parts of your body move as you breathe out?
- If you can, place one hand on your chest and place one hand on your belly. Which hand is your breath expanding into first? Are you breathing into your belly or are you breathing into your chest?
- Look down as you breathe in. Does your belly expand out or move back toward your spine?
- Don't change anything at all, just follow your breath.
- Are your shoulders moving up as you breathe in?

- Is there any tension in your body? Your face, your neck?
- Is there any tension in your mind?
- Is your mind busy or still?
- Can you feel your heartbeat as you breathe?
- Can you hear your breath flow in and out, or is it silent?
- If your mind starts to wander, that's OK, bring it back to your breath.
- Just be with your breath.
- What comes first? Your inhalation or your exhalation?
- Are you breathing fast? Are you breathing slow?
- Is your breathing deep? Or shallow?
- Is there a natural pause in between breaths or are you gasping for breath?
- Become aware of any changes in your breathing.
- Is it smooth and regular or a little jagged?
- Is your out-breath controlled, forced or relaxed?
- Gather as much information about how you breathe.
- Then come back into your space.
- Take a moment to write down anything you noticed about your breath, *e.g., nose/mouth, belly/chest, fast/slow, shallow/deep, its rhythm, its flow, etc.*

IAN

Once you become aware of your own breathing, you might start to become aware of how other people are breathing too. When Ian came through the doors of my breathwork studio in London, I could tell that he wasn't breathing in a functional way. But, as odd as it might sound, all I could really think of was Captain Cook.

I'd just got off a call with a friend who had recently moved to Hawaii. He excitedly told me the tale of Captain Cook and his crew's arrival on the island back in 1778, and how the Hawaiian islanders thought the reason for their visitors' pale skin was because they weren't breathing properly. They even changed the greeting we all know—"Aloha," meaning "the presence of breath"—to "Haole": *ha* means "breath," and *ole* means "no." "No breath" is a slang word locals use to describe foreigners (like my mate) to this day. Squeezed into white breeches and tight, gold-buttoned waistcoats fastened all the way up to the neck, Cook and his crew couldn't have been breathing comfortably, and the Hawaiians must have seen it. Now here was Ian, a senior director at a corporate organization, in his fitted suit, buttoned shirt and tie knotted tightly around his throat. I could already see he was very upright, rigid and tense. I could almost hear his mind whirring away, running up and down his to-do list as he came through the door.

"My wife told me you're the man to fix me," he said.

He didn't think anything was wrong with his breathing.

He explained he'd been sent to me by his wife because of his insomnia, but he was still highly functional. He was knocking off CrossFit sessions at 6:00 a.m., getting to his desk by 7:30 a.m., coffee in hand, spending the day in back-to-back meetings, smartwatch tracking his every move. But despite appearing to thrive in his daily life, he ended each day tired and increasingly low on energy and motivation. When he went to bed he'd stare at the ceiling for hours.

Bad breathing is like wearing your winter clothes on a hot summer's day.

FAR FROM FUNCTIONAL

It takes one to know one. I spent years like Ian—a Captain Cook. I led a fast-paced, goal-directed life, in a uniform—and my stress was made worse by that uniform, whether it was the suit and tie of corporate life or the skinny jeans I wore as a DJ. Many of us are just like me and Ian. And there are still plenty of people around who are blissfully unaware of how dysfunctional their breathing has become. Bad breathing is like wearing your winter clothes on a hot summer's day but not realizing it (and not knowing that your shorts are in your bag). It makes everything a little more difficult.

Although one in 10 people are reported to have "dysfunctional breathing,"[2] definitions of the term vary, and it's likely that the real figure is much higher. Experts say the prevalence of chronic rhinitis alone in the general population could be as high as 40 percent.[3] And before you say, "Rhinitis? That's a funny name, I don't have that," it might be worth knowing that rhinitis—inflammation inside the nasal cavity, often caused by a virus, allergens or dust—is known better by its key symptom: a blocked nose. Then add on the shallow breathers, the people who hold their breath when they type, those who breathe too fast or have erratic breathing, those who take periodic deep sighs and frequent yawns, as well as the chest breathers, or the ones with loud, heavy patterns, those who stop and start breathing in their sleep or who experience shortness of breath when playing sports…well, you get the idea. It's fair to say that, as a species, our breathing is far from functional.

Take a moment now to be aware of your breathing again. How are you breathing as you read? Often when we sit, our shoulders curl inward, our back rounds, restricting the flow of our breath in and out of the body. Is this you? If so, roll your shoulders back, straighten your spine, fix your posture, take a nice 'n' gentle, deep breath in, feeling your belly rise, and a slow, relaxed, long breath out, and carry on reading.

Despite the close relationship between the way we breathe and the way we think and feel throughout the day, few of us seek medical support for our breathing when we're anxious, lacking focus, low on energy or even experiencing more complex emotions, such as grief. We do not see breathing in the same way that we see chronic back pain, for example, or any other more obviously "physical" condition that has a negative impact on our lives. Unless you're experiencing a physical respiratory issue such as asthma, my bet is that you've never thought to visit your doctor to discuss your breathing, nor have you linked the way you're breathing to any mental or emotional complaint.

PUGS OF THE PRIMATE WORLD

You could make the argument that we're *all* dysfunctional breathers compared with our primate ancestors. As we evolved from ape to man, rather than evolving, our breathing *de*-volved. Two to three million years ago, our primate ancestors' brains grew, meaning that they needed extra space within the skull. So, like tectonic plates colliding, our faces shortened, our mouths shrank and our airways narrowed. Our nasal cavities were forced out into the open, creating our functionally compromised protruding nose, and our new, vertical air filter was exposed to more bacteria and airborne pathogens. This great step forward in the development of the human brain was accompanied by a step backward for our breathing function. Nature may have prioritized our mind, but it deprioritized one of the best means we have to keep it under control. It's like we became the pugs of the primate world.

AIR QUALITY

Our suboptimal snouts show their flaws when there's bacteria and pollution in the air we breathe. This is why, as air quality falls around the world as a result of the rise in toxic emissions and other human influences, more and more of us will suffer the symptoms of a blocked nose. This is, on the one hand, to protect our lungs from taking in dirty air, but the result is we end up breathing through our mouth with no protection. And most of us are affected: the Health Effects Institute said that in 2018, 95 percent of humans live somewhere where air pollution exceeds safe limits.[4] This is a real reminder to pay more attention to the impact the way you live may be having on the air around you. It's not just how you breathe, but *what* you breathe.

Our noses continued to evolve to adapt to different climates.[5] For those humans who settled in warm and humid tropical regions, noses become wider and flatter to help cool the surrounding air. Those who settled in Europe evolved much longer and narrower noses, to help to add heat and moisture to inhaled air.

Although modern science has now proved the Hawaiians and many other cultures right by showing us the importance of breathing, the relationship between breathing and good physical and mental health still goes largely unnoticed. Only a small, if growing, number of physicians are developing treatment plans around breathing, and there's a much wider societal lack of understanding about the influence of breathing on how we think and feel. The great thing is that this is beginning to change.

"BREATH IS LIFE"

In 1948 the United States Congress commissioned a study of the residents of the city of Framingham, Massachusetts.

The Framingham Heart Study, as it later became known, was a mammoth undertaking; it began with 5,209 men and women, and is now on its third generation of participants. Originally led by Dr. Thomas Royle Dawber, this ongoing study is responsible for discovering much of what we know today about heart disease. But what researchers have also found in the process is that the greatest indicator of life span isn't genes, diet or the amount of daily exercise we do. It's lung capacity and respiratory health. Researchers at the University of Buffalo, New York, came to a similar conclusion when they followed 1,195 men and women over 29 years with the aim of exploring the relationship between lung function and mortality.[6] They found that lung function is "a long-term predictor of overall survival rate in both sexes, and can be used as a tool in general health assessments." When I read science papers like this, I'm reminded of the Sanskrit proverb: "Breath is life, and if you breathe well you will live long on the earth."

Breath is life, and if you breathe well you will live long on the earth.

—Sanskrit Proverb

CHERRY ON TOP

Most of us are bad breathers, and it's affecting the way we think and feel for the worse. And here's the cherry on top: our breathing patterns are contagious.

Have you ever noticed yourself mirroring a friend's body language when you chat with them, perhaps by crossing your arms or legs to match them? This usually happens subconsciously or unconsciously, but you might have even noticed world leaders matching each other's body language consciously during meetings to help build rapport. Our breathing is no different. We often mirror the breathing patterns of others so that we bond with them. It's a nonverbal way of saying, "I feel the same as you," an involuntary expression of empathy. And since breathing affects how we feel, mirroring someone else's patterns means that we also meet that person at a similar emotional level. Mirroring breathing is another hardwired evolutionary survival tool—we need to be sensitive to the needs of our offspring and our species depends on group cooperation to survive.

You may have noticed this yourself, perhaps when speaking to a friend who has arrived late and flustered for dinner or coffee. You register her flushed face, her huffing and puffing, her erratic breathing. Unconsciously, your senses are altered. Your brain infers that whatever caused her stress represents a threat to you too. Your heart rate rises to match hers, your breathing pattern synchronizes. You meet her at her physiological and emotional level.

This also works in reverse. Have you ever been around someone who makes you feel calm, but you're not sure *why* you feel so Zen around them? Or someone who makes you feel pumped up and bouncing with positivity? It's probably because you've matched their energy and breathing patterns. Since breathing is contagious, you may choose to spend your time with people who share similar breathing patterns to you. Or perhaps you might do the opposite and feel drawn to people who breathe very differently to you, in the hope that you can disrupt your own perhaps dysfunctional breathing patterns and feel the way they do.

It's also very common for children to mimic their parents' dysfunctional breathing habits. Literally, "Like father, like son." In this way, one generation of bad breathers pass their bad breathing habits on to the next. And all this happens without your being consciously aware of it. But it goes both ways. When you master your breathing, you can be a positive example for other people and the next generation.

BEWARE OF THE NON-YAWNERS

Ever wondered why a yawn spreads through a room like wildfire? Like breathing, yawning is another echophenomenon. That's why you may need to watch out for those in the room that don't "catch the yawn" with you. A team of researchers at Baylor University recruited 135 students and measured their personalities for psychopathic traits before subjecting them to a contagious yawning experiment.[7] Guess what? Those who scored highly on the psychopathic scale were much less likely to yawn as they showed less empathy.

EXERCISE 3
FIND YOUR BREATHING ARCHETYPE

You can tell a lot about someone by the way they breathe. You'll instinctively notice a change in posture, facial expression and potentially even the changed breathing of a friend performing a stressful task, and know right away that something is up. You may not know exactly what, but you know intuitively from their breathing that something has happened. Every breathing pattern has an emotional signature, so if you understand the way breath flows (or isn't flowing), you can get a better idea of what you or someone else is feeling. Every breathing pattern—fast, slow, constricted, irregular—corresponds to a way of thinking, feeling and being.

Maybe you've noticed this yourself. You might have noticed your breathing change throughout the day when you've experienced different emotions. You might have recognized a change in breathing when you're excited, happy, sad, laughing or even crying. You might even have noticed that you hold your breath at times, maybe when concentrating on a text or email. In fact, we could say that studying your breathing pattern by bringing awareness to your breath, like we did at the start of this chapter in Exercise 2, is really like exploring a map: a representation of the physical, mental and emotional territory you occupy at any one moment in time. Your breathing reveals how your unconscious mind is perceiving your current world, and even maps out your past experience, by showing you, for example, what you're afraid of. Jane's fear of dogs was rendered visible in her style of breathing.

If you observe yourself and others around you breathing, you'll notice variations in depth, speed, length and location in the body. Have a go now. Try not to get arrested for voyeurism, but see what you can notice. Maybe you'll spot that someone is only breathing through their mouth. You might see someone holding their breath. Perhaps you'll notice someone's slouching posture really restricting how much their belly can move in and out. Maybe you'll spot someone looking super-Zen and relaxed, with a nice, open posture and their breath flowing easily in and out.

Breathing patterns vary from person to person. But although everyone's breathing is unique to them, there are common breathing styles that are easy to recognize. Posture tends to play a big part in nudging us into one of these categories. What follows is a description of some of the most common breathing patterns, and the things that cause each of them, something I was first introduced to by internationally renowned yoga instructor Donna Farhi, in her book *The Breathing Book: Good Health and Vitality Through Essential Breath Work*. These breathing archetypes all include causes for dysfunctional breathing that can be physical (posture, injury, illness) or linked to our emotions (stress, habitual thinking, past experiences). This might seem strange, but it's super-important that you recognize that your unconscious mind, shaped by your experiences, controls your breath.

Before reading the following descriptions, I'd like you to refer back to your answers to the breath-awareness exercise at the beginning of this chapter or pause and check your breathing again. Then, see if you can identify your breathing type and "breath-holding patterns." You may not fit neatly into one box, you may notice that you're a combination of two or more patterns or you may even recognize that your breathing fits into one of these archetypes before a certain event, such as a difficult conversation, a big performance or when you're deep in studying. With a bit more awareness, some of the breathing associated with these archetypes might be easy for you to fix. But if your pattern of breathing has been with you for a while, it will feel very "normal." We'll work through some quick fixes right away, and you will be able to release yourself from more stubborn breathing patterns as you work through the exercises in later chapters. But our first step is awareness.

So let's identify your archetype.

THE CHEST BREATHER

What is it? As the name suggests, chest breathing engages the small intercostal muscles in between the ribs instead of the diaphragm—the large dome-shaped muscle just below the rib cage. As these intercostal chest muscles are small, their short range of movement produces a shallower flow of breath. This increases the rate of breathing, to create a constant state of stress or anxiety. These small muscles also fatigue easily, which can limit your capacity for exercise.

Why does it happen? Chest breathing arises naturally when you're anxious, stressed or startled. Imagine once again that you've stepped off the pavement into the path of a cyclist—you gasp, and breathe high into the chest. That frightened breath is a chest breath. If you find yourself chest breathing by default, it might be a sign of habitual stress. It can also be linked to anxiety and to deeper sources of feeling inadequate, having poor self-esteem or deep-rooted fears. Working at a desk all day, wearing tight clothing or holding your stomach in to look slimmer can also cause habitual chest breathing.

What does it do? Chest breathing causes chronic tension in your upper back, shoulders and neck, which will return even after a massage for as long as you keep doing it. It can create chronic tension in your abdomen and cause the organs in the lower body to suffer from a lack of circulation. Because your primary diaphragmatic breathing muscle is not properly engaged, blood flow to the heart is impacted. Holding your stomach in to try to give the appearance of a slim waistline actually has the opposite effect: the tension restricts the circulation necessary for smooth digestion, while the disengagement of the proper breathing muscles hinders the assimilation of nutrients and elimination of waste products. This makes weight loss difficult, if not impossible. Psychologically, chest breathers may experience a busy mind that leads to anxiety and stress, which is probably the reason chest breathing has been linked to heart disease and high blood pressure. It's like the breathing you do when watching a nail-biting thriller—only all the time.

Test yourself: Place one hand on your belly and one hand on your chest. Breathe, feel and observe. Which moves more? Chest or belly? If your chest moves first, you're likely to be a chest breather. Chest breathing can also be accompanied by your shoulders moving up and down and a braced upper body.

The Chest Breather

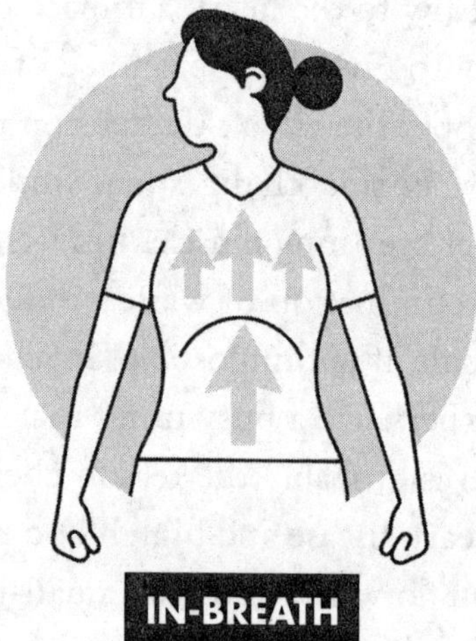

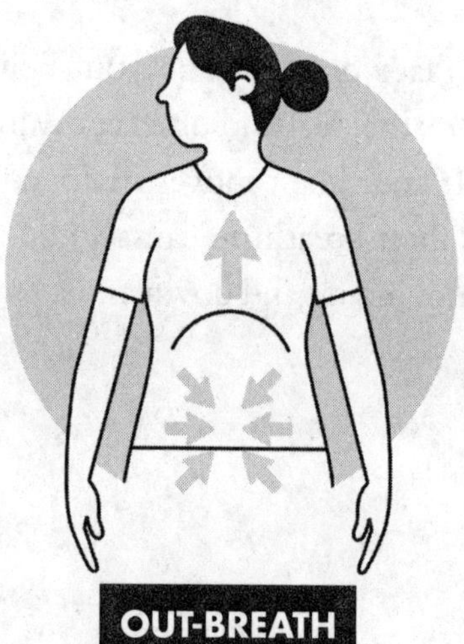

THE REVERSE BREATHER

What is it? In reverse breathing, the chest also expands before the belly when breathing in, but with this pattern the belly also sucks back toward the spine when you breathe in and then collapses outward when you breathe out. It's a kind of seesaw action between the chest and the belly, but in the opposite direction to what is natural, whereby the belly expands before your chest on the in-breath. In reverse breathing, your breathing motion is back to front: your abdomen never relaxes and your pelvic floor muscle contracts as you breathe in. Reverse breathing limits air flow into your body, leaving reverse breathers feeling as though they can't get enough air in through their nostrils. As a result, it's often accompanied by mouth breathing, which creates more stress.

Why does it happen? Restrictive clothing, such as tight belts and jeans, can be the culprit. It could also be due to a weak diaphragm, which, like any muscle, loses strength when it isn't used regularly. Or it could be due to stress, shock or fear, which paralyzes the diaphragm and prevents its natural movement. I've found there are often psychological reasons behind this pattern, often rooted in childhood experiences, which we will be exploring later on.

What does it do? Reverse breathers tend to have poor coordination, can be clumsy, and often seem stiff or awkward, especially when doing any kind of physical activity, such as dancing or sports. This is because one of their basic bodily movements—breathing—is, in a sense, upside down. Reverse breathing causes chronic tension in the upper back, neck and jaw. It can cause bloating, indigestion and heartburn, and might give you the feeling that you have a lump in your throat. Mentally, it can lead to disorientation and confusion. Extreme reverse breathers can seem spaced out and often find it hard to see the positive things in their life.

Test yourself: Reverse breathers often find it difficult to feel how their breath is flowing, so I'd like you to double-check and look at your body as you breathe. Place one hand on your abdomen and watch to see if your belly sucks in and moves back to your spine as you breathe in. Then, as you breathe out, do your shoulders and chest collapse and your belly push out? If so, you're reverse breathing.

The Reverse Breather

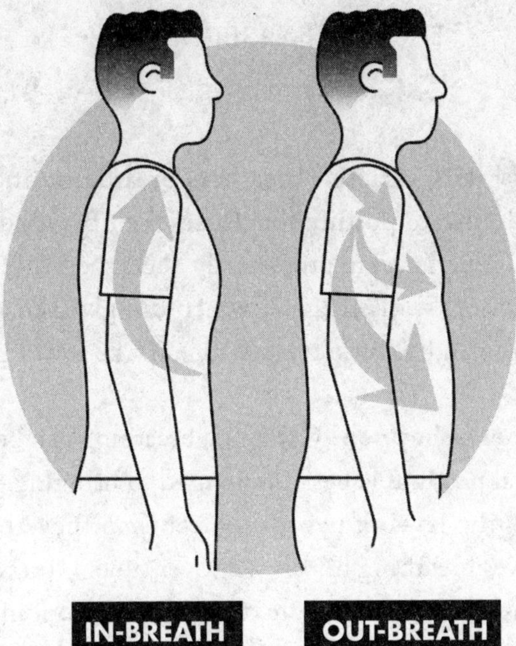

THE COLLAPSED BREATHER

What is it? Collapsed breathers breathe in and out in a hunched, inward posture that forces their breath downward. Their shoulders curl in protectively, their chest is drawn down and their core is disengaged, which can give the appearance of bloating and of little muscle tone in the lower body.

Why does it happen? Collapsed breathing can be the result of general postural issues. Those used to bringing themselves down to the level of people shorter than they are can find themselves breathing in this way over time. Obsessive smartphone users can develop "text neck" from looking down at their screens all the time, as can sitting at a desk and computer or driving all day. Your posture may collapse over time, causing you to develop this breathing pattern. Collapsed breathing can also be a result of negative past experiences. You collapse your chest and shoulders in as a means to protect yourself from emotional pain. In some cases, collapsed breathing may also be the result of trauma or abuse; the mind develops a survival technique by disconnecting itself from sensations below the neck. Collapsed breathing can be a consequence of issues with body image and can develop in people who grow up feeling disconnected to their body—like a "head on legs."

What does it do? There are two types of collapsed breather: some collapsed breathers sigh and gasp, often in an effort to get more air into their lungs. They frequently experience tiredness, headaches and shortness of breath. Mentally and emotionally, they can feel down, emotionally closed and pessimistic. In some cases, however, collapsed breathers can be lively and animated, but only from the neck up. These individuals have a poor sense of their physicality and often live mostly in their mind, occupying a world of ideas in which the body seems of little use, unable to reflect the animation in their eyes and face. Both types of collapsed breathers have limited blood flow into the heart and lungs, which can create health complications.

Test yourself: Collapse your chest, curl your shoulders down and let your belly protrude. Does this feel familiar? Posture is something we do not always pay attention to, so ask those around you if they notice you curling your shoulders inward or hunching over. If people are often telling you to sit or stand up straight, then this is likely to be an archetype you frequently experience.

The Collapsed Breather

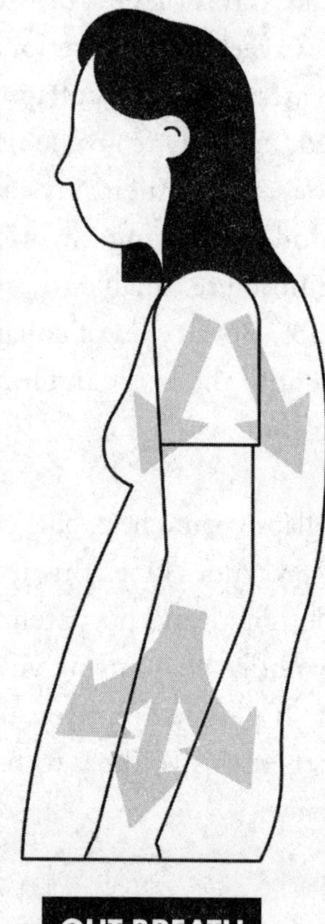

THE FROZEN BREATHER

What is it? Frozen breathers have a tendency to halt the movement of their breath and restrict its flow in and out. Imagine your posture when you cringe at something or being outside on a winter's day without a jacket. This is similar to the frozen-breathing posture, as frozen breathers tend to restrict their breathing and hold their breath unconsciously. In healthy breathing there is a very short, natural pause between inhalation or exhalation. Think of it as a micro-rest between breathing cycles. If this pause unconsciously lasts longer than one second, then your breath could be frozen.

Why does it happen? Frozen breathing can happen after a period of rapid shallow breaths (which are usually stress-related). These rapid shallow breaths reduce the amount of carbon dioxide in your body, so it halts breathing to let it attain balance; a sigh will start the cycle over again. Many people, especially goal-oriented types, can also find themselves entering this pattern in an "I'll breathe when I'm done" mentality. They might, for example, stop breathing until they hit "send" on an email. Chronic frozen breathing may also be a result of deep-rooted fear—the fear of being seen or heard, the fear of moving forward in life, the fear of letting go of the past. In extreme cases, the freezing of the body and breath can be another unconscious safety mechanism to cope with overwhelming feelings. I often see this type of breathing with people suffering from PTSD.

What does it do? Frozen breathing upsets the chemical balance in your body, which in turn may create an erratic breathing pattern—fast breathing, no breathing, periodic big sighs—on repeat. It's as if you're playing catch-up for missing breaths. Frozen breathing can reduce your energy levels and hinder your ability to work. Mentally, involuntary breath holding can warp your perception of time. Chronic breath holders can be unemotional and rigid.

Test yourself: Be aware of your body. Is your breathing almost undetectable? See if you can tell if there's air flowing in and out. Are you contracting or tensing your muscles, as you might if you were cold? Next time you're working on a difficult or time-sensitive project, pay attention to your breath. Do you find yourself "forgetting" to breathe, maybe when sending an email or an important text?

The Frozen Breather

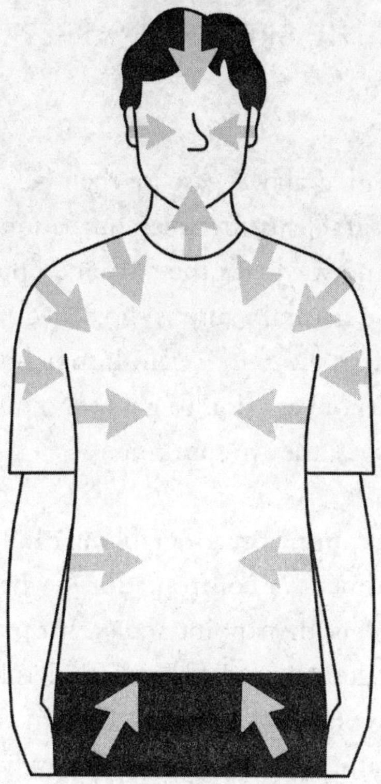

THE BREATH GRABBER

What is it? Breath grabbers gasp for their next breath without allowing the natural pause between inhalation and exhalation to occur. In some ways, it's the extreme opposite of frozen breathing, as the natural pause is shortened. Chronic breath grabbers will often jump in to finish someone's sentence for them, cut into a conversation to get their point across or run out of breath when they're speaking.

Why does it happen? Breath grabbing can happen in times of conflict, argument or confrontation—whenever someone is rushing in to get their point across. It can also happen in times of real excitement, when someone has lots of ideas and finds it hard to contain themselves. Chronic breath grabbers may also have a real need to prove themselves, be heard or become the center of attention; they're always trying to voice their thoughts and opinions, moving quickly from one thing to the next. A childhood experience might have cemented the belief that if they don't jump in, they'll be left out or left behind, which leads to this breathing pattern. It can also be an avoidance of being fully present.

What does it do? Breath grabbers tend to develop a faster breathing pattern and, by extension, higher blood pressure. They often talk fast and loud, and can run out of breath when speaking. As this pattern doesn't allow our body and mind to rest between breathing cycles, breath grabbers can feel busy, rushed and pressured. Someone with this breathing pattern may find it hard to listen and tends to feel uncomfortable with silence and stillness. They may find it hard to relax and accept life as it happens.

Test yourself: Monitor your breathing during your next conversation. Do you butt in? Do you want to tell your story? Notice whether you allow yourself to pause, whether you allow the other person to complete their sentence. Do you find yourself "running out of breath" while speaking or giving presentations at work?

THE BREATH CONTROLLER

What is it? The breath controller's breathing is fairly functional. Throughout this book, you will be controlling your breath, because by doing so you can control the way you think and feel. But when I talk about the breath-controller archetype, I'm talking about people who habitually control their out-breath. We only want to control our out-breath when we need to, not all the time. The controller pattern is harder to spot because they look like healthy breathers. Believe it or not, I see this pattern a lot in yogis and athletes, where conscious control is a big part of the activity. And through so much practice, this becomes their default day-to-day. It's also common in managers and senior executives, who are responsible for many people and different streams of work.

Why does it happen? I often call this the "super-achiever" type, because controlling breathers tend to exert a lot of influence on their surroundings, and have a strong sense of their own personal agency and responsibility. Controlling breathers often think that they're right, and "my way or the highway" can be their default. People with this pattern can have difficulty trusting others due to past experiences, which is why they try to control all their circumstances. Of course, trying to control all your circumstances is like trying to control the weather. You simply can't.

What does it do? Breath controllers reinforce their lack of trust that things can't be left alone without needing to be influenced or controlled, and this pattern can make it increasingly hard for them to let go of thoughts and feelings. The breath controller will also feel they always have to be doing something and find it hard to simply be. They can be vulnerable to major shocks or catastrophes, because they can't control them, meaning that major events can feel overwhelming. In extreme cases, the breath controller can develop fears of activities where they have little or no control, such as flying.

Test yourself: Where is your breath flowing? Does your breath appear to start in the midsection between your chest and belly? Does this midsection part of your torso appear to rise before anything else? Now investigate your out-breath. Is it controlled or forced? Imagine breathing in is like pulling an elastic band and breathing out is letting go of it, so the band springs back into place by itself with no effort or need for control. Does your out-breath spring back effortlessly, or are you forcing or controlling it? Perhaps you're already thinking you have to get this right? If so, then you may be the breath-controller archetype.

THE PERFECT BREATHER

What is it? Sadly, there's no such thing as one perfect way to breathe, as how you breathe depends entirely on what you're doing. However, when I look for the perfect breath, I like to start with the natural resting breath. When you're resting, your breath shouldn't be stressed, tense, fast or restricted. The ideal resting breath flows in through the nose, and begins in your lower torso like a wave rolling upward. It's expansive, yet smooth, slow, relaxed and effortless. A perfect breather will have a very slight pause between breaths. They exhale through their nose and their breathing is relaxed. They exert no control or force.

Why does it happen? Healthy breathing reflects a healthy mind and a healthy body. Perfect breathers may not have avoided any misfortune or difficulty in their lives (because that's impossible), but they have processed any difficult emotions and learned to let go of the things holding them back.

What does it do? Perfect breathing provides the optimum flow of oxygen to your cells, for whatever task is being performed. At rest, perfect breathing keeps your whole system balanced and helps eliminate toxins from the body. Perfect breathers feel focused, relaxed and joyful. They're able to go with the flow but can also take control of things when they need to, managing not to be overcome by their emotions in stressful situations. This kind of breathing creates a feeling of security, optimism and contentedness. When you breathe in this way, you're smiling at life and life is smiling back.

Test yourself: The perfect breath when resting will not be the perfect breath for you for all moments in your day. As you continue to read this book, start to notice your breathing in these five key areas in your day and make a note about which archetype best represents each.

1. Breathing at rest
2. Breathing during your day
3. Breathing when sleeping
4. Breathing when speaking
5. Breathing during physical exercise

BREATHING'S ARCH ENEMIES

I spent my judo years holding my abs in, my corporate years hunched at a desk in a suit and tie, my DJ years in tight skinny jeans. I could barely breathe. Consider what your ability to breathe fully will be when faced with the following arch enemies of breathing. Addressing any of these can also work as some quick wins to ease the grip of a number of the negative breathing archetypes.

- Clothes that are too small or too tight—if you can't run or dance in it, you can't breathe in it! We all know what it's like to feel held to ransom by beauty standards or the latest fashion trend, but to breathe properly, you need to choose clothes that you can move around in. So get your comfies on.
- Correct your posture—are you a serial sloucher? Have you got text neck? Do you stoop down to talk to people shorter than you or look up to speak to people taller than you? Bring some conscious awareness to your posture, because it has a huge impact on your breathing flow. Remember to always stand up straight with your shoulders back (even when texting your mates).
- All-day desk sitting—if you're sitting hunched in front of a screen all day, it's difficult to breathe properly. Make sure

you sit correctly: spine straight, uncross your legs, feet flat on the floor—and take breaks to move and stretch.

- Belts—if it pinches when you sit down, it's too tight and will limit the natural movement of your diaphragm. Be mindful that your belt might feel a little tighter after a big meal, which will affect your digestion. Loosen the belt or ditch it altogether.
- Ties and button-up shirts—OK, some dress codes and special occasions will require you to wear a collar. But if you must wear one, make sure it's loose so there's no tension in your throat.
- High heels—again, save these for special occasions. They move your center of balance, which makes your breathing muscles contract and tighten. This restricts the flow of air in and out.
- Bras—it's been widely reported that many women—some say up to 81 percent—are wearing the wrong size! If it leaves red marks, it's too tight. Switch it.

3

SHUT YOUR MOUTH AND SLOW THINGS DOWN

SHUT YOUR MOUTH!

Seriously, shut it. Keep it closed for the rest of this book—and forevermore, for that matter. Tape it up if you have to. No, I'm not joking. We'll get to mouth-taping very soon. I know it may be challenging at first, but if you can, one of the quickest ways to fix your breathing is to breathe through your nose.

Mouth breathing naturally occurs when you're under stress; it helps ring the alarm bell and trigger your sympathetic system's fight-or-flight response. So yes, sometimes your mouth breath *will* save your life—like from those cyclists and bears—but if you're habitually breathing through your mouth, day and night, stress will overwhelm your body and mind. The sympathetic side of the nervous system will dominate, your heart rate will go up, your blood pressure will rise, stressful thoughts will become your default. There are a number of reasons for this, which we'll be getting into later on in this chapter, but for now, just keep that mouth closed.

It's estimated that 30–50 percent of adults breathe through the mouth, especially during the early-morning hours.[8] If the side effects of habitual mouth breathing were on the side of a bottle of air, they'd read something like this:

WARNING: May cause asthma, high blood pressure, poor digestion, heart disease, allergies, decreased lung function, exhaustion, poor sleep, snoring, bad breath, sleep apnea, dental decay, gum disease, dysfunction of the jaw joint, narrowing of the dental arch, jaw and palate, constriction of the airways, crowded and crooked teeth, loss of lip tone, noisy eating, poor speech, swallowing problems, trauma to soft tissues in the airway.

Is your mouth still closed? I hope so.

FACE SHAPE

When you breathe through your mouth, an external force is exerted on both your upper and lower jaw, so the muscles in your cheeks have to work harder and become taut. The more frequently you breathe through your mouth, the greater the influence of these forces, which over time can narrow the shape of the face as well as the dental arches. Narrow face and dental arches mean less room for your tongue—which drops down to the floor of your mouth, instead of resting against its roof. This also makes your tongue more susceptible to sliding back into your airways when you lie down or sleep, which can obstruct them and cause you to stop breathing—a sleep breathing disorder called sleep apnea.

The tongue dropping to the floor of the mouth, which happens when you mouth breathe, has been shown to hinder the development of the face in children.

It's been suggested that because we no longer need to chew hard, as our ancestors did, our jaws don't get the workout they need to grow. This results in a smaller nasal passage and a propensity to mouth breathe. But by chewing dental gum, alongside nose breathing, you can strengthen your jawbone, and carve out space in your airways.

30–50 percent of adults breathe through the mouth, especially during the early-morning hours.

EXERCISE 4
NOSE-UNBLOCKING TECHNIQUE

You may be reading this chapter thinking, “Hold up, Stu. I can’t breathe through my nose, it’s proper blocked up.” If you’re just experiencing temporary stuffiness, which could be due to allergies, a cold or anything that has you bunged up, this exercise is better than any nasal spray I’ve ever come across. It works by temporarily increasing the level of carbon dioxide in your body, forcing your body to clear blocked airways to allow more oxygen in. Moving your head shifts the mucus caught in your nose.

- Take a normal breath in through your nose. (If your nose is too blocked for that, breathe in through one side of your mouth, like Popeye.)
- Take a normal breath out through your nose (or through one side of your mouth).
- Gently pinch your nose and hold your breath.
- Tilt your head slowly to the left.
- Tilt your head slowly to the right.
- Tilt your head back.
- Tilt your head forward.
- Come back to a neutral position.
- Breathe in through your nose gently.
- Breathe out through your nose.

- Repeat this as many times as necessary until your nasal airway is clear. (If you're gasping for air on the in-breath, then you've held it for too long.)

If you're still finding it hard to breathe through your nose, it may be due to a structural issue. If this is the case, I'd advise that you get checked by your doctor.

Take a normal breath in, and a normal breath out of your nose.

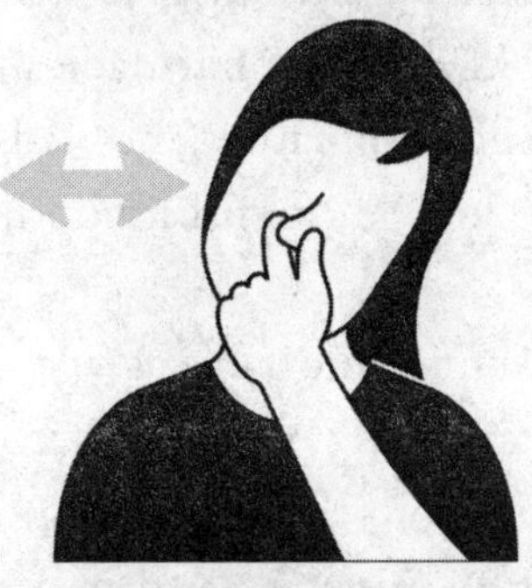

Pinch your nose and tilt your head to the left, then the right.

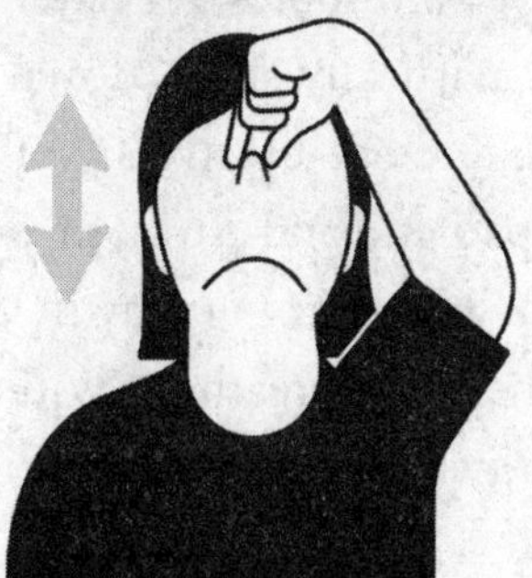

Keep your nose pinched and tilt your head backward, then forward.

EXERCISE 5

TAPE IT SHUT

With your nose now nice and clear, I want you to learn to nose breathe at all times. The quickest and most effective way to do this often raises some eyebrows at first. But trust me on this one. It's been tried and tested, time and again. It's simple, it's effective, and it helps air flow the way it should and balance out your whole system.

It's time to tape your mouth shut.

I know this may sound pretty scary, but we're going to take baby steps.

Do not use duct tape, packing tape or anything crazy sticky. Use micropore medical tape, sold at most pharmacies. It's soft, easy to pop open if you need to breathe (and, lads, it doesn't rip out your facial hair!). Just be careful if you have chapped lips.

The mouth tape may feel a little uncomfortable at first, and if you feel anxious about it, just start with it on for 30 seconds a day. When that feels easy, build from there until you can have the tape on comfortably for 20 minutes per day.

Breathing through the nose naturally slows down your breathing so that your in-breaths and out-breaths are of equal lengths, balancing out the sympathetic and parasympathetic sides of your nervous system. We'll be learning why in Chapter 4. But for now, just know that nose breathing will allow you to better manage your energy, thoughts and feelings.

KNOCKED TOO MANY TIMES

"Dude, it's like I'm breathing through a crooked straw when I inhale through my nose. Do you know how many times I've bashed this thing?"

This was the first session with a martial arts client of mine, who competes in UFC. The moment he opened his mouth, I could tell he was having difficulty nose breathing by the sound of his voice. Your nose is the narrowest part of your respiratory tract, so nose breathing can feel almost impossible if the airway is blocked. This can be due to structural collapse of the septum, the thin wall between the nostrils. A deviated septum is fairly common in contact sports but can happen in any accident that impacts the nose. (One of the most common ways is, bizarrely, walking into a wall.) Some people are also born with deviated septums. I'd love to say I've a magic wand to treat them, but they require surgery. If you get to the end of this chapter and think this might be you, have a chat with your GP.

THE COOLEST CLUB IN TOWN

The first time I went to Berlin I was told about a legendary bouncer named Sven Marquardt. One of the most famous doormen in the world, he's the tattooed, stone-faced gatekeeper of the Berlin nightclub Berghain, a coveted venue notorious for its inscrutable door policy.

At Berghain there are no reservations. There's no bottle service. Cameras are "verboten" and even the lenses on smartphones are taped over.

There's no way to get on the guest list—unless you're one of the DJs playing there. Many people spend hours in the queue, only for Sven to politely turn them away without explanation. Sven also happens to be the only bouncer in the world to have done a menswear collaboration with Hugo Boss.

Your lungs are like Berghain, the coolest club in town. And your nose is Sven Marquardt.

But why does our nose only want to let certain "guests" into the lungs?

Well, the air you breathe isn't just made up of those pollutants we read about in Chapter 2. It can include many other things—foreign particles, dust, pollen, germs, viruses, mold, organic compounds, metals and small particles, all of which irritate your lungs. To protect you from these, your nose filters, purifies and sterilizes the air. It gets it in shape before it's allowed into the club.

It does this in a number of ways.

The first line of defense are the small hairs inside your nose that immediately filter out large particles. These hairs are equivalent to the person on Sven's team who scans the queue while you wait to get inside.

Next is the mucous membrane, a moist, thin layer of tissue lining the inside of your nose, which makes mucus, that sticky, snotty stuff.

The mucus is like Sven's guy who pats you down and checks your pockets or bag. It picks up dust, debris, bacteria, fungi—anything floating in the air that could cause upset to the body. If something does get trapped in there, it's either broken down by the body's defenses, ends up blown on a tissue or—ACHOO!—you sneeze it out. It doesn't get into the club.

Once inhaled air has swooped into your nose and the small hairs have filtered out any larger particles, it proceeds to three seashell-like structures called turbinates. As the name suggests, they stir and circulate the air, and heat, moisturize and humidify it. Your lungs and throat do not tolerate air that's too dry, cold or dirty. So this burly bouncer for the lungs screens the air molecules waiting in the queue to go inside, letting some in and turning others away. This process means that air is delivered to your lungs at a perfect temperature and moistness, preventing damage to the delicate tissues in the lungs and optimizing the function and delivery of oxygen.

A healthy nose can efficiently provide about 90 percent of the heat and moisture required to condition inhaled air.[9] The nose is a two-way street—when air leaves your body as you breathe out, your turbinates help to prevent any loss of heat and moisture from the body. Studies have shown the nose retains 33 percent of exhaled heat and moisture.[10]

So when you breathe in, your nose captures anything we don't want in the body—including viruses. On the out-breath, the warmer air kills these off and sends them out of the nose.

As you breathe in, the network of turbinates in your nose also helps to create resistance, which controls the flow of air into your body.

You might think that more resistance sounds bad; why not bypass the nose and breathe through the mouth, to get more air in? However, the increased resistance that comes from nose breathing allows the right amount of pressure to build in the lungs, which improves the efficiency of your breathing.

By nose breathing, you create about 50 percent more resistance to the air you're taking in than you would if you were mouth breathing, and this enables you to inhale about 10 to 20 percent more oxygen.[11] This improves blood oxygen circulation as well as carbon dioxide levels, so that your system is more balanced. You also need sufficient nasal resistance when you take a breath in order to maintain the elasticity of your lungs.

In a nutshell, simply by breathing through your nose, you optimize the air arriving in your lungs and slow your breathing down to a rate where your body and mind can perform at their best.

Let's take a moment to pause and let the wondrous nose work its magic.

EXERCISE 6

LET THE NOSE DO ITS THING

- If you have some micropore tape, pop a piece over your mouth.
- Breathe in through your nose for 1, 2, 3, 4, 5.
- Let your nose clean, moisturize and warm the air.
- And breathe out: 1, 2, 3, 4, 5.
- Let your system balance and your mind become calm.
- Breathe in: 1, 2, 3, 4, 5.
- And breathe out: 1, 2, 3, 4, 5.
- Once again…
- Breathe in: 1, 2, 3, 4, 5.
- And breathe out: 1, 2, 3, 4, 5.

THE MIRACLE MOLECULE

The nose is a magical thing. But it has one more superpower worth exploring.

When you breathe in through your nose, your sinuses release a colorless gas called nitric oxide into the inhaled air, which is then transported to the lungs. Nitric oxide was considered poisonous until 1998, when three American scientists, Robert F. Furchgott, Louis J. Ignarro and Ferid Murad, discovered that it was "a signaling molecule in the cardiovascular system."[12] Ignarro would go on to call it the "miracle molecule."[13] They won the Nobel Prize in medicine for their work.

But what does "miracle molecule" mean? What these scientists discovered is that nitric oxide has the incredible ability to make your blood-vessel walls relax and widen. This creates more room for your blood cells to travel through them, thereby increasing blood flow and lowering blood pressure. This process is called vasodilation, and it enables blood, nutrients and oxygen to travel to every part of your body, effectively increasing your oxygen capacity. Nitric oxide has also been shown to act as a bronchodilator, which means it relaxes the muscles in your lungs and widens your airways.[14] This means air can easily get where it needs to go.

Nitric oxide is antifungal, antiviral and antibacterial. It protects you from flu, pneumonia and a range of other viral infections. During the SARS outbreak and the COVID-19 pandemic, nitric oxide therapy was tried out with success on patients, shortening hospital rates and lowering the rate of fatality.[15, 16] One key reason for the results was that inflammation in the patients' lungs decreased.

NASAL CYCLE

Try this.

- Gently close one nostril with your thumb.
- Give a short, sharp inhalation through the open nostril, but not so much that your nose closes.
- Repeat this on the other side.

You should be able to feel which nostril is more open. You can also listen to distinguish which nostril has a higher pitch. The nostril with the lower pitch is more open.

If you practice this throughout the day, you'll notice the airflow through one nostril is more dominant than through the other, and that it appears to switch. This is known as the nasal cycle. It occurs naturally, and the length of time between the switch happening differs from person to person, with the length of the average cycle ranging from 30 minutes to two hours.

According to yogic tradition, the nasal cycle is related to brain and nervous system function. When one nostril is dominant, the opposite hemisphere is more active. So when breathing is dominant in your right nostril, the left side of your brain is more active; when your left nostril is dominant, the right side of your brain is more active. Although both sides of the brain are involved in everything you do, each side is believed to have a bigger role in carrying out specific tasks. Your right nostril links to the sympathetic branch of the ANS, increasing its activity, while your left nostril links to the parasympathetic branch of the ANS, increasing its activity.[17] Nasal dominance also affects the autonomic nervous system.

Legend has it that some yogis only make decisions when both nostrils are equally open and their ANS completely balanced, on and off in equal measures. This would certainly limit the number of decisions you can make in a day!

You can start to balance them out through a simple practice called alternate-nostril breathing.

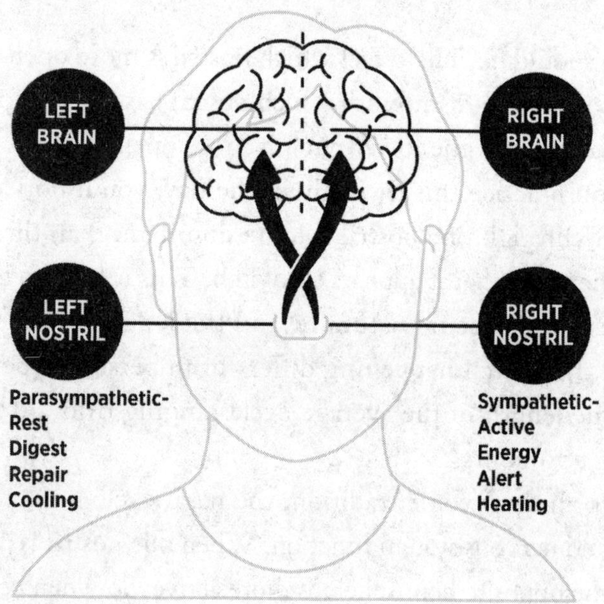

A KINKY SECRET

Ten healthy subjects took part in a study that measured levels of nitric oxide when humming compared with a quiet exhalation at a fixed flow rate. They found that the oscillating airflow caused by humming on the nasal cavity increases nitric oxide levels 15 times more than during quiet exhalation.[18]

As nitric oxide is an effective way to enhance blood flow, some medications, including Viagra, harness the nitric oxide pathway to promote blood-vessel widening and improve blood flow to the penis to enhance erections! Ladies and gents, you have a natural Viagra just a few hums away.

EXERCISE 7
ALTERNATE-NOSTRIL BREATHING

This practice has balanced in-breaths and out-breaths, and you switch which nostril you breathe in and out of. Switching nostrils balances out the two sides of the autonomic nervous system and the different sides of the brain. When working with kids, I sometimes call alternate-nostril breathing "mountain breath," as you breathe in going up one side of the mountain and breathe out as if climbing down the other. You then climb back up the second side and back down the first to where you began.

Let's give this a go, step by step.

- Make a peace sign with your right hand. (We will be using the thumb and ring finger.)
- Close your right nostril with your right thumb, and breathe in through your left nostril steadily for a count of four.
- Pause as you close your left nostril with your right ring finger and open your right nostril.
- Breathe out calmly through your right nostril for a count of four, and pause briefly at the end of your out-breath.
- Breathe in through your right nostril steadily for a count of four.
- Pause as you close your right nostril with your right thumb and open your left nostril.

- Breathe out calmly through your left nostril for a count of four.
- Repeat.

One cycle will help, if that's all the time you have, but to maximize the benefits, I'd aim to practice for four minutes or more to balance the hemispheres of your brain, as well as your sympathetic and parasympathetic responses.

Make a peace sign with your right hand.

Close your right nostril with your right thumb and breathe in through your left nostril for a count of four.

Close your left nostril with your right ring finger and breathe out through your right nostril for a count of four.

BREATHE IN, BELLY OUT

Now that your mouth is closed, the air is flowing nicely in and out of your nose, and both nostrils are balanced, there's another key player in breathing well that I want you to meet. You met briefly in the archetypes section, as it's impossible to discuss breathing types without mentioning it: your diaphragm. Your primary breathing muscle.

The diaphragm might be the most underappreciated muscle in the human body. It really doesn't get much street cred, maybe because when it's flexed your belly sticks out. Only mammals have this muscle, and it has been speculated that its evolution, which gave mammals a highly efficient way of taking in a steady supply of oxygen, made it possible for our ancestors to evolve a warm-blooded metabolism. In fact, without the diaphragm, humans might not have been able to evolve big, powerful brains that require a lot of oxygen.

So the diaphragm matters. It's vital for healthy breathing, correct posture and organ function. It also supports both the vascular (or circulatory) system, made up of the vessels that carry blood through your body, and the lymphatic system, the main function of which is to clear toxins from your cells. If you don't use your diaphragm, it tightens up and you'll soon start to breathe like the chest-breather or reverse-breather archetypes. In fact, engaging our diaphragm and breathing by using it correctly could go a long way toward fixing almost *all* our dysfunctional breathing archetypes.

So what is it? Well, the diaphragm is a massive sheet of fibrous muscle that acts like a divider, separating your chest, lungs and heart from your abdominal cavities, which contains part of your digestive tract, the liver and pancreas, the spleen, kidneys and adrenal glands. Think of it this way: if your rib cage were a birdcage, only encasing the heart and lungs, the diaphragm would be the base of that cage, running across the whole cross-section of your body.

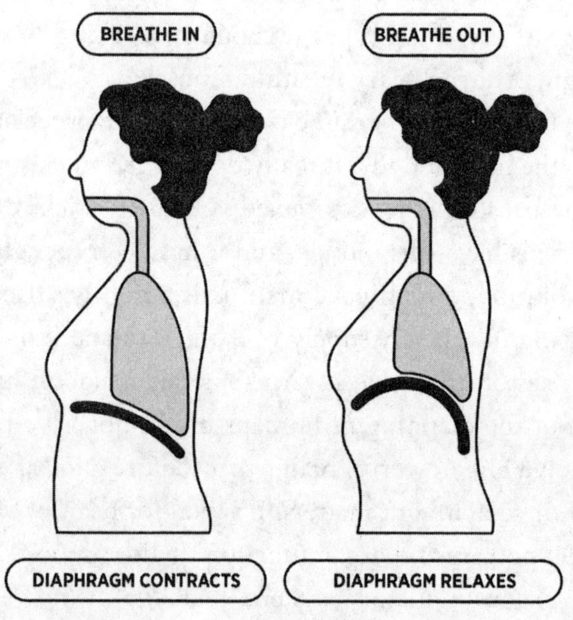

In a natural, relaxed in-breath, your diaphragm contracts and opens downward toward the base of your spine, a bit like a parachute. This movement displaces air and organs downward, creating a 360-degree expansion—your lower belly expands outward first, followed by your side ribs and midsection, and finally your intercostal muscles (the ones in between the ribs in your chest) contract and your lungs expand. This increases the space in your chest cavity, increases lung volume and decreases lung pressure so air is drawn in. You'll notice that your belly moves out before there's any movement in your chest. As everything moves downward, the lower torso expands, which is why diaphragmatic breathing is often referred to as abdominal breathing or belly breathing.

I try to avoid using the terms "belly breathing" and "abdominal breathing," as both are a little misleading. It's the displacement of the air in your lungs and your organs moving down that cause the belly to expand. You're not using or engaging your stomach muscles in this process. They remain relaxed.

As your diaphragm moves down, it massages your internal organs—stomach, small intestine, liver, pancreas and the kidneys. Your kidneys will move two to three centimeters with a proper diaphragmatic breath and your organs get squeezed like sponges. This helps massage organs, move fluid around, increases nutrition delivered to your cells and aids digestion. This movement also acts as a pump for your lymphatic system, which removes toxins and protects your body from illness. Your diaphragm is attached to your heart via a sac called the pericardium. This envelops the heart, so as you breathe your diaphragm and heart move together, increasing blood flow and helping your heart perform more efficiently.

Your chest muscles, shoulders, neck and secondary muscles also have a role to play as they contract and expand, but in a resting breathing pattern, these are secondary, less active and only get to work after the primary muscles have been engaged. They may become more active in times of stress or danger to get air in and out quickly, but hopefully this shouldn't happen too often. When we talk about breathing, the diaphragm is king. And if you're not using it, it can cause more issues than you might think.

When we talk about breathing, the diaphragm is king.

EXERCISE 8

RELEASE DIAPHRAGM TENSION

Like any muscle, the diaphragm can get tense. The most obvious way to notice this is limited movement in the lower part of the torso when you breathe in, like our chest-, reverse-, collapsed- and frozen-breather archetypes. Here's how you can release it and help it regain its full range of movement.

- Run the fingers of your right hand down your breastbone until you reach the end of your sternum. (Watch out for the xiphoid process, the little bit of delicate cartilage at the bottom.)
- With your fingers here in your solar plexus area, slide them to the right, to the upper base of your rib cage.
- Apply light pressure, by hooking round the bottom edge of your rib cage.
- Breathe in deeply, maintaining the light pressure.
- Breathe out.
- Move your hand farther down your rib cage.
- Breathe in.
- Breathe out.
- Work all the way down one side.
- What do you feel?

- If you find a spot of tension (and you'll know!), stay there for a couple of breaths to see if it eases.
- This can feel painful, but it shouldn't be a sharp, shooting pain. It should feel more like a knot or point of muscle stiffness.
- Now repeat down the left side with your left hand.

For many, this exercise is uncomfortable because the diaphragm is tense and tight. Every time you get a shock, a fright or a moment of stress, the diaphragm draws in tightly. If you live in a large city, you may have got used to the tension that comes whenever you hear a siren wailing outside your window. But the more you learn to breathe properly with your diaphragm, the freer it will become. You'll feel a fuller range of movement and any tension will be released.

Run the fingers of your right hand down your breastbone.

Slide your fingers to the right, then hook them round the bottom edge of your rib cage. Breathe in, then out.

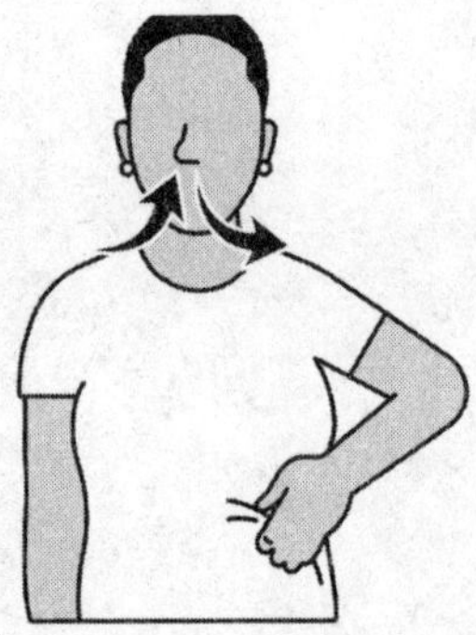

Continue sliding your hand farther down your rib cage, pausing to breathe in, then out.

Having loosened some of the tension in your diaphragm with Exercise 8, let's stretch things out and improve your range of movement.

EXERCISE 9

PUSH THE SKY

I love this exercise. Like a morning stretch, it helps to lengthen and expand the diaphragm's range of motion, as well as those of the surrounding intercostal, oblique and abdominal muscles. At the same time, it helps to correct your posture so that you breathe efficiently and effortlessly. This is really important if your breathing falls into the collapsed- or frozen-breather archetypes.

- Extend both your arms out in front of you with the palms facing each other.
- Let the middle fingers of your hands touch.
- Now turn your palms down toward the floor and let your elbows face out, and with your middle fingers still touching, push down toward the floor until your arms are straight.
- Breathe in through your nose, feeling your belly rise while lifting your hands above your head. Keep your arms straight, middle fingers touching, palms end facing away from you up to the sky.
- Hold your breath and push the sky for four seconds. Feel the stretch across your midsection and diaphragm. Don't push from your shoulders.
- Breathe out. Circle your arms round and down by your side.
- Repeat five rounds.

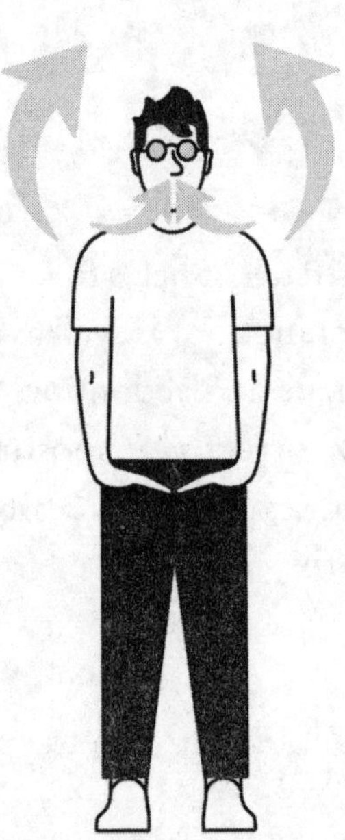

Breathe in through your nose and lift your arms above your head.

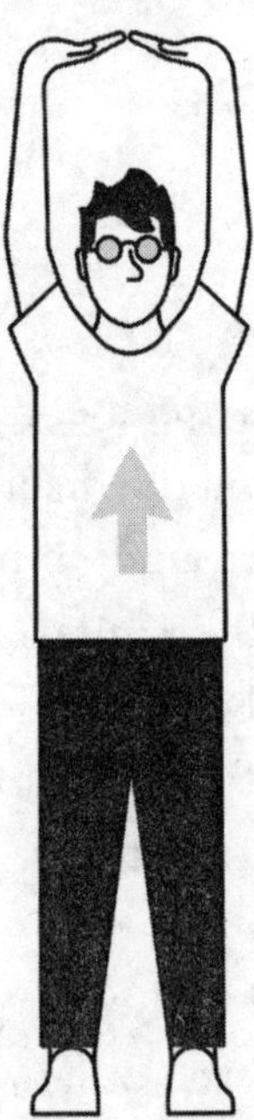

Hold your breath and push the sky for four seconds. When you breathe out, circle your arms round and down by your side.

EXERCISE 10

DIAPHRAGMATIC BREATH

Now you have worked out some tension and stretched things out, let's practice some deep diaphragmatic breaths.

- Sit in a comfortable position or lie flat on the floor.
- Relax your shoulders.
- Put both hands on your belly.
- Breathe in through your nose into the hands on your belly. The displacement of the air and organs downward will make your stomach rise before the chest. (Try not to push out with your stomach muscles. By tensing your abs and pushing them outward, you can make your belly stick out, but the diaphragm isn't engaged or working when you do this.)
- Breathe out through your nose, letting your belly fall back to where it began.
- Repeat for 10 rounds.

If you find this easy, try to keep breathing through your diaphragm as you continue to read.

If you're really struggling to feel your belly rise and fall, you could try this lying flat on your front. In this position, you should be able to feel your belly moving against the floor.

Sit in a relaxed position with your hands on your belly.

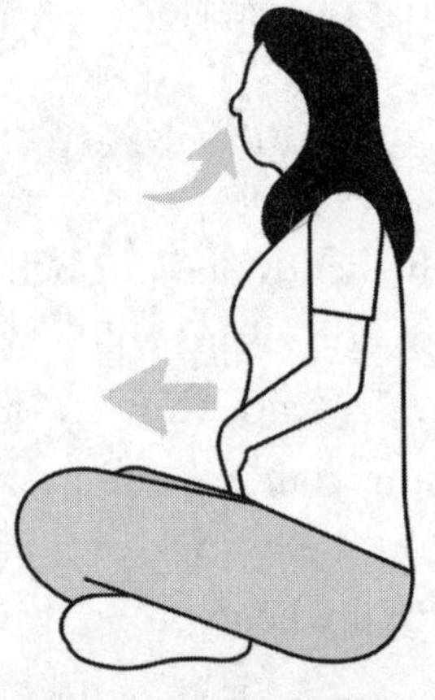

Breathe in through your nose and feel your belly rise.

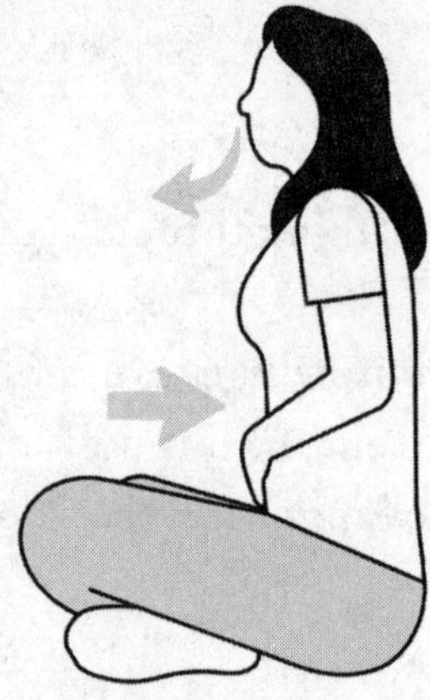

Breathe out through your nose and feel your belly fall.

You've now met breathing's key players: your nose and diaphragm. I want you to use them as much as you can to breathe. Come back to Exercises 8 (Release Diaphragm Tension), 9 (Push the Sky) and 10 (Diaphragmatic Breath) each day to work out any tension in your diaphragm and get used to using it. Also, stop and pause several times a day to check if you're still using your nose and diaphragm. They're the mechanics that keep the machine that is you running smoothly, letting humid, filtered, heated air into and out of your lungs optimally, supporting almost every function in your mind and body—even your posture and face shape. They keep your energy in balance and your mind settled, helping you to maintain an equal flow of breath in and out.

Let's explore this further.

DON'T WASTE YOUR BREATH

"You may or may not read this straightaway," said Ali, handing me the book. "It will find you when you're meant to read it."

Ali is my best mate, a brother from another mother. Our friendship formed during a summer season in Ibiza when we were 18. He's quite possibly the most positive person I know and I trust his opinion. He wasn't usually this cryptic with a gift.

I'd later learn that the book he gave me was not your usual book. Paramahansa Yogananda's *Autobiography of a Yogi* was not the kind of book I was used to reading. It wasn't inviting. The orange front cover looked dated, with a picture of a long-haired yogi. When I opened it on my way home, I discovered the text was tiny, the pages thin. The book looked daunting—it was full of yogic terms—so it went on the shelf and I stopped thinking about it.

Three years later, I was reading by chance about Steve Jobs. I learned that while he was battling the cancer that eventually took his life, he planned every aspect of his own memorial service. Held at Stanford University in 2011, the attendees included the former US president Bill Clinton, Microsoft founder Bill Gates—Jobs's longtime rival—and John Lasseter, the former chief creative officer at Pixar. To my surprise, I also learned that each of his guests had received a small brown box containing *Autobiography of a Yogi*. It turned out that Jobs first came across the book when he was 17, and he reportedly read it every year. It was the only book he kept on his iPad. I also found that it's listed at the Institute of Advanced Study in Princeton, New Jersey, as one of Albert Einstein's favorite books.

All of a sudden, it found me. It came off the shelf. I soon discovered that the book, which I found challenging to read at first, describes Yogananda's "self-actualization" and is full of timeless wisdom. One of Yogananda's insights is that the faster an animal breathes, the shorter its life span. "The restless monkey breathes at the rate of 32 times a minute," writes Yogananda, "in contrast to man's average of 18 times. The elephant, tortoise, snake and other creatures noted for their longevity have a respiratory rate that's less than man."[19] Dogs, cats and mice have a high respiration rate and live a markedly shorter time than a giant tortoise, for example, which takes only about four breaths per minute and can live up to 200 years.

Yogananda drew much of this wisdom from the Vedas, the oldest scriptures in the Hindu tradition, which also state that the number of breaths allotted to us for our lifetime is predetermined when we're born. The quicker we spend our breaths, the sooner we die—which gives a whole new meaning to the phrase "wasting your breath."

MAGIC RATIO

Just as you can only take in a certain number of calories before you gain weight, there are limits on how often you can breathe and how much air you can take in before you experience negative effects. "Healthy" breathers have a respiratory rate of 12 to 18 breaths per minute and take in around five to eight liters of air every 60 seconds. Breathing any more than this is hyperventilation, a massive no-no. It's like adding too much fuel to a fire—it gets out of control. Even if you take in just a little more than your allowance every day, it does not take long before your system goes out of balance. Your sympathetic nervous system becomes overly dominant, which is exhausting. It feels like you're running from a grizzly bear all day, every day.

Just as the tortoise breathes slowly and lives long, there are clear benefits to bringing your breathing down below these parameters. It's something yogis, meditators and Eastern spiritual teachers have practiced for thousands of years. Many in the West have experienced it through yoga and other practices that have risen in popularity over the last century. It's only in the last decade, however, that researchers in the West have started to study breathing in depth and find a scientific basis for what has been known anecdotally for millennia. It turns out the Eastern traditions were right all along.

Researchers have explored the potential benefits of slow breathing on the cardiovascular, nervous, respiratory, endocrine and brain systems.[20] They've discovered that our optimal resting breath rate is almost half the number set in healthy breathing parameters, snailing in at between five and six breaths per minute—that's roughly five seconds in and five out, with a slight pause between each cycle.[21] Funnily enough, this pace matches the breathing rate yogis maintain when reciting mantras or prayers. Researchers in Italy have found that "rhythm formulas that involve breathing at six breaths per minute," such as praying while using the rosary, induce powerful psychological and possibly physiological effects.[22] Even singing in a choir helps to slow down the rate at which you breathe to near these numbers.[23] After all, your voice is simply a sounding breath.

When we think of breathing, we think of the lungs, but the heart also has a big role to play. Breathing in a coherent way, with our magic ratio of in for five and out for five, mimics this smooth heart pattern and helps control your heart rate variability (HRV), a measure of the variation in time between each heartbeat. A more controlled HRV helps to improve cognitive function, and reinforce positive feelings and emotional stability.

EXERCISE 11

MAGIC RATIO BREATHING

- Sit in a comfortable position or lie flat on the floor.
- Relax your shoulders.
- Put both hands on your belly.
- Breathe in through your nose into your hands on your belly for a count of five.
- Breathe out through your nose for a count of five.
- Repeat for 10 rounds or more.

See if you can breathe at this magic ratio for the rest of this chapter and be more like a tortoise. If you find it challenging and start to feel like you're lacking in air, you might be surprised to hear that this feeling of "air hunger" has nothing to do with oxygen, and has everything to do with carbon dioxide.

AN IMAGE PROBLEM

Breath is air we take in and breathe out again. But what's in the air? Mostly oxygen, nitrogen, carbon dioxide and hydrogen. And everyone knows that oxygen is good and carbon dioxide is bad. Right?

Not so fast. Carbon dioxide isn't just breathing's waste product. It's a vital part of the process. But for some reason it's got itself a bad rep. Everywhere we look, its name is tarnished. Carbon dioxide is the bad guy, and the best thing for us to do is to get rid of it. But it's the balance of carbon dioxide and oxygen that's the backbone of life on earth. Without it, we wouldn't be here. Nothing living would be here. Carbon compounds regulate the earth's temperature, make up the food that sustains us and provide us with the energy that fuels our global economy.

Carbon dioxide traps heat close to the earth, helping the planet hold on to the energy it receives from the sun so it doesn't escape back into space. If it weren't for carbon dioxide, the earth's oceans would be frozen solid.

Oxygen is pretty much the new kid on the block when it comes to our atmosphere. It has been relatively scarce for much of our planet's 4.6-billion-year existence. Life on earth was initially pretty boring. But around 2.5 billion years ago, something interesting happened. A big blob of sludgy, bluey-green algae called cyanobacteria started a trend. The bacteria drank some water, soaked up some rays and started to breathe in carbon dioxide to produce energy. We call this process *photosynthesis*.

Cyanobacteria produced another gas, called oxygen. This was of no use to them, so they spit it out into the oceans and air. Cyanobacteria had a right old party with carbon dioxide, and as a result, the oxygen content of the atmosphere rapidly increased. Earth underwent what American geologist Preston Cloud first described in the 1970s as the Great Oxidation Event (GOE). The buildup of oxygen caused by the GOE started to react with the sun's ultraviolet light and formed a protective ozone layer that shielded the surface of the earth from too many harmful solar rays. Oxygen, being a reactive gas—literally a fire-starter—became a new source of energy for the other life-forms now able to survive on the surface of the oceans and on land.

By breathing oxygen, organisms could move beyond the simple life-form of cyanobacteria, and become bigger, more active and more intricate. They became animals, from worms to fish to mammals. The earth as we now know it began to teem with life, and all because of this life-sustaining symbiotic balance of oxygen and carbon dioxide. Interestingly enough, shipping cyanobacteria, or blue-green algae, to Mars to re-create the GOE and make Mars humanly habitable has already been successfully tried and tested in the lab environment, where the atmosphere of Mars was replicated.

So carbon dioxide is not the enemy. Life on planet earth needs it. Without both oxygen and carbon dioxide, life doesn't exist. And it's the same with breathing. We don't just need oxygen over carbon dioxide. We need both. Let me tell you why…

ALL ABOARD THE HEMOGLOBIN BUS

Getting oxygen into your lungs is simple: you take a breath. But getting oxygen from the lungs into your cells is a more complex task that requires the right balance of carbon dioxide. As you breathe in, oxygen from the air around you spirals down your windpipe to your bronchial tubes, the tubes that connect your windpipe to your lungs. When it reaches your lungs, it enters the alveoli (air sacs), where oxygen is passed into your bloodstream. In your bloodstream, it's pumped by your heart and transported to your cells by a protein called hemoglobin, which is found inside red blood cells. Hemoglobin won't off-load oxygen to the cells unless carbon dioxide is present in the cell. This means that if carbon dioxide levels are too low in the cell, oxygen will not be delivered to it.

Think of it this way: on its way to work, oxygen travels from the air to the lungs; here it jumps aboard the hemoglobin bus, passing through the heart and on to its desired cell destination. The hemoglobin bus will drive right by oxygen's bus stop unless carbon dioxide is waiting there. However, if carbon dioxide is there, it can hop on, swap seats with oxygen, and ride the bus to the heart and then the lungs to be breathed out. So balancing oxygen and carbon dioxide is very important, and although carbon dioxide is often dismissed as a "waste product," it's a vital part of respiration.

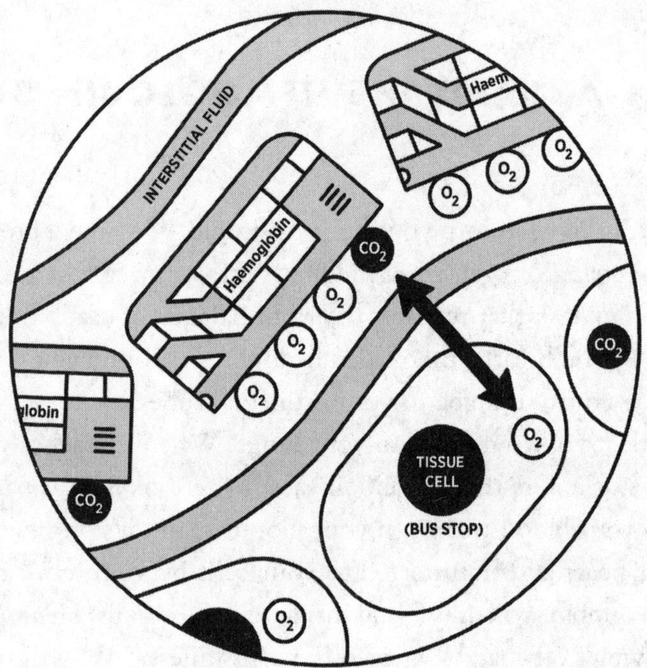
INTERSTITIAL FLUID
Haem
Haemoglobin
globin
CO_2
O_2
TISSUE CELL
(BUS STOP)

TIP THE BALANCE

You might have noticed when breathing through your nose, practicing our magic ratio, or even breathing with your diaphragm, that you feel as though you're not getting enough air.

But this desire to breathe, and even the uncomfortable feeling of breathlessness when we're "hungry for air," isn't caused by a lack of oxygen, as you might imagine.

In fact, your brain's immediate trigger to breathe is caused by an increase in carbon dioxide. The reason for this is that the body wants to keep your pH balanced, and when carbon dioxide is in water it becomes carbonic acid. This means that when you hold your breath, or slow your breath down, the amount of carbon dioxide within your blood increases, your pH level drops and your blood becomes acidic. Sensors in your brain detect this and ring the respiratory alarm bell, telling you to take a breath. This happens before lack of oxygen is a problem.

Your body requires you to stay within a tight range of pH—between 7.35 to 7.45—and if it strays out of that range, there are consequences for your digestive system, immune system, nervous system, muscles, joints and skin, to name a few. You're more prone to mood swings, colds and flus, nausea, arthritis, and breakouts of spots and blemishes. So your body works hard to keep your pH balanced in that sweet spot.

TAKE YOUR BREATH AWAY

Spanish architect Aleix Segura i Vendrell knows a thing or two about the uncomfortable feeling of not taking a breath. In 2016 he broke the Guinness World Record for holding his breath for a whopping twenty-four minutes and three seconds. This is not something he did without practice. As a child, he learned to free dive (dive underwater without breathing apparatus) in the sea during family vacations on the Costa Brava.

The breath-hold times that Aleix and many free divers can achieve sound impossible to non–free divers. Lung size and genes certainly play a part. But it's something you can train to improve, and the science behind this superpower is actually quite straightforward. You can undergo training designed to increase your lung capacity and carbon dioxide tolerance, as well as your ability to work with lower levels of oxygen. We will explore how to do this in Chapter 9. So you'll be (not) breathing like Aleix in no time.

HYPERVENTILATION, CARBON DIOXIDE AND PH

The short, shallow, rapid breathing associated with stress, which we've explored in other chapters, alongside the over-breathing that's associated with mouth breathing, changes the pH level in your body. Faster breathing or over-breathing, known as hyperventilation, decreases the partial pressure of carbon dioxide as you expel more air. Less carbon dioxide in your bloodstream means less acidity, so your blood becomes more alkaline, something called respiratory alkalosis. During moments of acute stress, your body can cope fine with this brief change in pH. But if a stressful day becomes a week, a month, a year, then your body will need to do something about the prolonged change in its pH levels caused by this habitual, rapid, stressful breathing.

Now, lab coats at the ready while I explain. We're going to get a bit science-y for a moment. When you breathe too much over a prolonged time, in an attempt to keep within the healthy pH range, your kidneys expel fewer hydrogen ions in your urine, and decrease bicarbonate reabsorption and production. This is a fancy way of saying that your body holds on to more acidity to balance your pH levels. But when the kidneys do this, your body's chemistry finds a new normal breathing rate. Once your pH has been reset in this way, the slightest increase in carbon dioxide will throw you out of balance again, and you'll feel the desire to breathe soon after your last breath. The result? You need to keep breathing at this faster pace to feel like you're getting enough air, thereby becoming trapped in a state of hyperventilation. You stress more, so you breathe more, so carbon dioxide decreases, so your kidneys make alterations so that your pH remains in the healthy range. Your pH might balance out, but now you're stuck in a stressful over-breathing pattern, sensitive to the slightest change in carbon dioxide. Let's do a quick diagnostic to check if this is something affecting you.

EXERCISE 12

CARBON DIOXIDE TOLERANCE TEST

This simple test was first developed by the Russian scientist Dr. Konstantin Buteyko, a former expert mechanic who, after the Second World War, decided to ditch vehicles and instead research the "most complex machine, the Man." He said, "I thought if I learned [about] him, I'd be able to diagnose his diseases as easily as I had diagnosed machine disorders."[24] He went on to develop methods of treating asthma and a wide range of breathing-related ailments. This test, which he called the "Controlled Pause Test," was designed to test carbon dioxide tolerance and highlight any issues of hyperventilation and its negative side effects.

In the following exercise, I want you to hold your breath and count how many seconds it takes until you feel the *first desire* to breathe. This is not a maximum breath hold. We're looking to identify the first urge to breathe—it may be a thought or a feeling. You can use your phone or a stopwatch to time yourself if you like to be more accurate.

- Take a normal breath in (don't be tempted to take a large breath, as this may skew your score slightly).
- And a normal breath out. Relax and let go.
- Pinch your nose to hold your breath and begin counting.
- Stop counting when you feel your very first desire to breathe.

- Take a note of your score and return to normal breathing.
- Repeat this each week. Preferably first thing in the morning.

Scoring, explained:

- **Below 12 seconds:** highly sensitive to carbon dioxide. Correlates with an increased breathing rate. More likely to experience more stress, anxiety and panic.
- **12–20 seconds:** mild sensitivity to carbon dioxide. A vulnerable system easily prone to experience stress and anxiety.
- **20–30 seconds:** good tolerance of carbon dioxide. A calmer, more stable, more resilient system.
- **30–40 seconds:** low carbon dioxide sensitivity. Correlates with slow breathing and good physical, mental and emotional health. Likely to feel more relaxed and calm.

EXERCISE 13

SLOWLY DOES IT

If you scored below 20 seconds on the carbon dioxide tolerance test, slow diaphragmatic, nose breathing will help improve it. It's a good one to practice, even if you scored high. The idea here is to slow your breath right down, so that you experience a mild feeling of air hunger. I want you to teeter on the edge of being comfortable to help reset any hyperventilation patterns. This forces your body to create more red blood cells and improve the delivery of oxygen to your cells, meaning that your breathing becomes more economical. Oxygen uptake increases, you harness nitric oxide and therefore increase your carbon dioxide tolerance. And here's a bonus: it anchors your mind and helps you concentrate too.

Set a timer for five minutes daily. You could do this when you have your mouth tape on, if you like. We will start off with our magic ratio, then slowly increase the in-breath and out-breath lengths so that your breath slows down even more.

- Using your nose and diaphragm, start with our magic ratio breath.
- Breathe in for a count of five, feeling your belly rise.
- Breathe out for a count of five, relaxing your body as you feel your belly fall.
- Repeat.

- If you're feeling comfortable, increase the count by one. So…
- Breathe in for a count of six, feeling your belly rise.
- Breathe out for a count of six, feeling your belly fall.
- When this is comfortable, increase the counts further to in for seven, out for seven.
- Try to find the edge of your air hunger. Don't push over that edge. Work with it.
- Repeat and keep increasing the numbers each day until you can do a complete five minutes breathing in for ten counts and out for ten counts.

In this chapter we've explored the miraculous power of the nose and diaphragm, and how slowing the rate at which you breathe to our magic ratio of five seconds in and five seconds out or even slower can transform your physical and mental well-being. We've also spoken about the importance of carbon dioxide, and how, by testing your carbon dioxide tolerance each week, you'll be able to see how your breathing is improving as you continue on through this book. Now, of course life happens, things get stressful sometimes, so now that you're aware of more functional breathing, I want to show you how to turn your stress to calm.

4

STRESS LESS, SLEEP BETTER AND MANAGE PAIN

FROM STRESS TO CALM

The alarm you've come to hate jolts you out of a pleasant dream. You're groggy. You roll over. You hit snooze and it goes off again, just as you fall back asleep. Already you're an anxious mess.

Soon your mind is a swirling vortex of to-dos, should-dos, must-dos. Your thoughts are in overdrive. Your heart pounds. Your breathing becomes short and shallow. Stress hormones flood your body. Your day feels overwhelming. And you haven't even had your breakfast yet.

No time for breakfast. Negative thoughts fill your headspace. *I should have got up earlier. I never do anything right. I'm a failure. I look like shit today.* Your breathing turns erratic.

And now your phone's in your hand. Your inbox is overflowing. Your mind is scattered across 10 apps, 20 conversations. You try to do a million things at once.

The day is a constant stream of rings, dings and pings. Something takes you by surprise and your breathing is now faster than Usain Bolt's Pumas. You find some time to focus, then realize you're forgetting to breathe altogether. You find moments to gasp, sigh, yawn, vent. By the end of the day, you collapse into the sofa at home, exhausted. You snore through the latest Netflix series. When you finally make it to the bedroom, you spend what should be your sleeping hours staring at the ceiling, thinking about the next day.

Does all this sound familiar to you? Stress and anxiety are stark realities of modern life. In 2018, in a Mental Health Foundation survey titled "Stressed Nation," it was reported that 74 percent of UK adults had been so stressed over the course of the previous year that they felt "overwhelmed" or "unable to cope."[25] It's the same story in the US; Everyday Health researchers said in 2019 that chronic stress was a "national epidemic for all genders and ages."[26] Both these studies were conducted *before* the COVID-19 pandemic. If you navigated that period without any stress, please do get in touch, because I'd like to know your secret!

This stress epidemic can be made worse by major events—be they global or in your personal life—but the truth is that stress has its roots in the nature of modern living, from our "always-on" lifestyles, to our addiction to smartphones, to political volatility, wars, or even the climate crisis. There are also parts of the world affected by poverty, famine, disease. Stress is everywhere.

As we get accustomed to stress, it disguises itself in the very fabric of our experience. It takes the controls without you noticing, and it does this so well that you may even be unaware that your body and brain have been hijacked. You might feel like you're just built this way.

As we explored in the previous chapter, when you have a stressful breathing pattern over a long period of time, your body literally adapts itself into a new normal that then forces you to continue breathing in this way. It makes you become stuck in a particular mode of breathing and experience the stressful feelings this produces. You may even find it hard to think or act differently. It's a vicious circle—you become so accustomed to a certain amount of stress physically and mentally that your brain and body register it as normal. In order to maintain the same stress level, you require more and more stress over time.

The worst part is that this doesn't just happen on a physiological level; you also start to become *addicted* to the thoughts and behaviors that produce more stress. You might become so addicted that you unconsciously wait until the very last minute to get something done, or take on more projects than you can actually handle. You might even choose a relationship full of drama just so you can feel like your familiar stressed self. Or you might simply ask for a triple shot in your coffee to raise your cortisol to a level that feels "normal."

In this chapter I'm going to show you that stress doesn't have to define your life. When things are overwhelming us, they get mixed up with our emotions and often seem much worse than they are; but if we take them out of our heads, they lose their power, and we can also start to think about how we can break them down and deal with them.

Have a think now.

- When do you feel anxious or stressed?
- Before a social event or maybe a big meeting?
- Perhaps before you have to speak in front of a group of people?
- What about when you're running late?
- Is a state of stress or anxiety something you feel is your "normal" state of being?
- Do you tell people "I'm just an anxious person"?
- I'd also like you to see if you can recall how you breathe in the situations where you feel stressed or anxious.
- What do you notice about your breath?

As we explored in Chapter 1, when you're in a state of stress and anxiety, the sympathetic side of your ANS has been activated. Now that you understand how this and its opposite, the parasympathetic system, work, in this chapter I'm going to show you how to use your breath to take back control of your stress, dial it all the way down to calm and regain harmony in your mind and body.

Stress doesn't have to define your life.

IF IN DOUBT...

When it comes to turning stress to calm, we first need to understand our breathing is quite binary: your in-breath, however it comes in, will switch you on to some degree, engaging your sympathetic drive (your stress response). Your out-breath switches you off—the parasympathetic takes over, and your body and mind relax. So by playing around with the speed, ratio and lengths of in-breaths and out-breaths, you can control which side is more dominant.

Let's say you're in a waiting room before a big job interview. You've spent the whole week preparing. You couldn't *be* any more prepared. But as the time of your interview approaches, you start to feel restless and fidgety. Nerves kick in. You start chest breathing, your hands clam up, you feel flushed, your mind starts going on overdrive… And then the voice in your head pipes up.

I really need this job. Wait—what if I don't get it? I'm screwed if I don't get it. And if I don't get it, I'm probably not good enough. I'm definitely not good enough. How am I even here in the first place? I bet I'm not as qualified as anyone else. Shit, were these even the right shoes? I never do anything right… And why is it so hot?

Now your breath is frozen. Your mouth dries up. Your hands tremble.

Life gets kinda stressful sometimes, and turning that sympathetic stress into parasympathetic calm all starts with a phrase. These might be the six most important words in this book. Stick 'em on your fridge, tattoo them on your arm—whatever you do, don't forget them. They've saved my bacon too many times to remember:

IF IN DOUBT,
BREATHE IT OUT.

If in doubt: if you're feeling anxious, worried, overwhelmed or panicked, if you can't sleep, are struggling with digestion or feel in pain, are nauseous or even struggling with motion sickness, really, anything that's causing you to feel out of sorts, then *breathe it out*. A prolonged out-breath activates your parasympathetic response, promotes rest, digest and repair. If you calm your breath, your mind will follow.

EXERCISE 14

IF IN DOUBT, BREATHE IT OUT

- Breathe in through your nose for a count of four, feeling your belly rise.
- Hold your breath for a count of four, keeping calm and still.
- Breathe out through your mouth for a count of eight. Drop your shoulders, relax your face, your jaw, behind your eyes, even your forehead.
- Repeat as much as you need until you feel relaxed.

When you're in a stressed or anxious state, your thoughts tend to accelerate and your inner voice can get louder or begin to catastrophize, like in our interview example, and this can keep you hostage to your stress response. So it may take a few rounds of our "If in Doubt, Breathe It Out" technique to calm yourself. You may wish to add some positive self-talk too.

POSITIVE PEP TALK[27]

We can replace the negative talk in our brains that often accompanies stress by consciously engaging in positive self-talk. When your brain says, "I can't do this, I'm useless," you can say instead, out loud or in your head, "I've got this, I can totally do this, I'm doing the best I can." In fact, over time and with practice, positive self-talk can replace negative self-talk, so in stressful situations you're more likely to respond with encouragement. You may want to go back to your intentions and add or change one of your statements into an example of positive self-talk that you can engage in throughout the day.

You can also start the morning by getting yourself into a positive mindset, so your day runs that little bit smoother itself. Every morning ask yourself:

1 What would make today great?

2 What am I grateful for?

When you think about the positive, you learn to be more optimistic. It's a positive intention for your day ahead. And gratitude is great anytime stress kicks in: it's a positive emotion that we feel when we receive something, and it causes our brain to flood our body with feel-good chemicals, making it impossible to feel stress or fear. Give it a go now. What are you grateful for right now?

VAGUS NOT VEGAS

When it comes to hitting the off button, you best get friendly with your vagus nerve. Don't let the name throw you: although this nerve influences your breathing and heartbeat, makes you sweat and gets you talking, it's not to be mistaken for the highly stimulating party town Las Vegas. The vagus nerve is the opposite; it's a main player in your parasympathetic response, which makes you feel relaxed. It has two divisions: ventral vagal (rest and digest) and dorsal vagal (freeze). This nerve acts like the eyes for your internal system, knowing if something is out of balance.

- In the ear, it has the sense of touch, letting you know if there's something lurking within.
- In the throat, it influences your vocal cords, enabling you to speak.
- It influences the muscles at the back of the throat responsible for the gag reflex, if something's stuck there.
- It influences breathing and helps your heart rate to slow.
- It influences the contraction of muscles in the gut to aid digestion.

Thanks to the vagus, if you're hungry you'll know to eat, if your bladder is full it will stimulate the muscle that helps you pee, if you're stressed it can press the brake pedal to slow down your breathing, and lower your heart rate and blood pressure.

In Latin, *vagus* means "wandering," so it's a fitting name, being one of the longest cranial nerves, stretching from our brain right down to our torso, connecting major organs on its way. It's a whole network—a pathway, an information highway—made of many nerves that links your ears, throat, lungs, heart, stomach, liver, spleen, kidneys and intestines. It's like a phone hotline between the organs and the brain, enabling the brain to keep track of what's happening in the body. And it's the reason behind your mental state influencing your digestion, heart rate and breathing.

Some scientists like to measure how "on" the vagus nerve is as vagal tone. It can be used as an indicator of how stressed you are. Someone with high vagal tone has better control of their body's stress response compared with someone with low vagal tone. It's even said that vagal tone is passed from mother to child, so if a mother is stressed and has a low vagal tone, this will likely be the same for their child.

The vagus is an important component in the slowing down of your heart rate when you breathe out. Your organs also send messages to the brain via the vagus. Despite being part of the parasympathetic side of the nervous system, the vagus isn't always linked to calm states. Branches of the vagus can be quite stimulating as well. They can even make you vomit. So maybe it's a bit of Vegas after all.

How can we use the vagus nerve to feel calmer? Through breathing. Next time your unconscious mind triggers a stress response—you feel your heart rate increasing and your breathing speeding up—slow your breathing. If in doubt, breathe it out. In for four, hold for four, and breathe out of your mouth for eight. This sends a signal to your other organs and your brain that there's no threat. Everything is cool and you start to feel calm. So anytime stress or anxiety kicks in—before a work meeting, a date, an exam or for no apparent reason at all—you can breathe your way to feeling more chilled.

The vagus is like a phone hotline between the organs and the brain.

POLYVAGAL THEORY

According to behavioral scientist Stephen Porges, the body doesn't just use its sympathetic and parasympathetic systems in isolation; we're not just "on" or "off." Instead, Porges believes there are times where our body is in a hybrid state of activation and calming, and using both our sympathetic and parasympathetic systems. He calls this state our "social engagement system," and believes that it's connected to our vagus nerve, and that it plays a large part in our ability to engage socially.[28]

Social engagement requires a calm, settled openness that only can happen in a safe "ventral vagal" state, which correlates with the parasympathetic side of the nervous system and the rest and digest response. Any stress, anxiety, worry or sympathetic arousal from this state creates action—"I can." But if the arousal is too much or too sudden, we can go back into a parasympathetic state, only this time, we enter the dorsal vagal state, and "freeze"—"I can't." We withdraw, shut down, feel hopeless, even trapped—as in the case of people in deep depression. Porges proposes that you have to pass through the sympathetic state to move from the hyperarousal dorsal vagal back to the safety of the ventral vagal, just as you have to pass through it to go the other way. This is something that can be difficult for someone who is depressed, numb, overwhelmed, or feeling acute helplessness or hopelessness, to achieve. The theory can be hard to get your head around, so spend some time looking at the diagram.

It has been hypothesized that breathwork techniques that arouse a sympathetic response in a controlled way can enable someone to feel their way back to a healthy state of "social engagement," defined by openness, compassion, mindfulness and groundedness. We will practice some of these techniques in Part 2: Deeper Work (page 199).

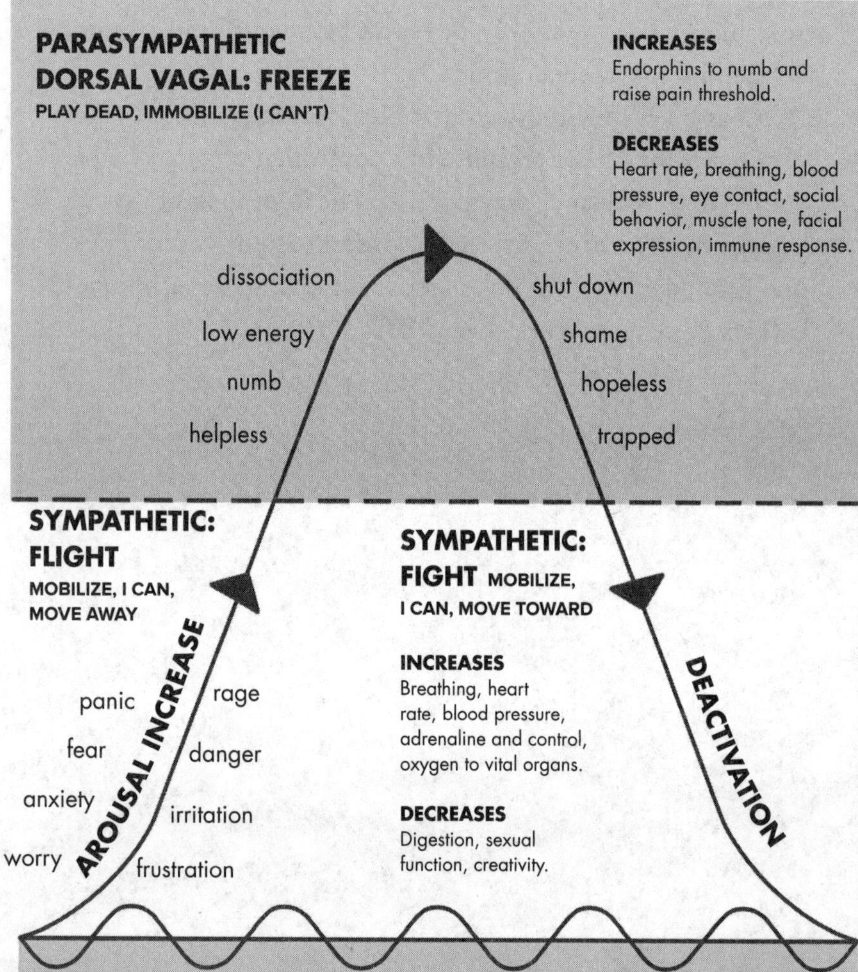

TO MOVE FROM A PARASYMPATHETIC DORSAL VAGAL STATE (YOUR FREEZE RESPONSE) TO A PARASYMPATHETIC VENTRAL VAGAL STATE (A PLACE OF SAFETY), YOU MUST PASS BACK THROUGH A SYMPATHETIC (STRESS) STATE.

TAKE TIME TO CHILL

Our "If in Doubt, Breathe It Out" technique can extinguish the fire of stress or panic in most situations. There are, however, a few other handy tools you can harness to improve your rest and recovery, especially for sleep and digestion, by provoking a parasympathetic response. The great thing is that a good night's sleep and healthy digestion can also stop us from getting too stressed in the first place.

First, though, I'd like you to take a moment to think about those times when you feel most calm and relaxed. Most of us can recognize the activities that reliably land us in our ventral vagal state and make us feel relaxed, open, grounded and safe. Any kind of self-care works: taking a hot bath, lying on the sofa, lighting a candle or stretching, for example. But in the chaos of our lives, we often don't set aside enough time in our day to do them. Taking time to chill is so important. Think about scheduling time for rest, as well as work or any other more demanding activities. Stop doing, just be.

When do you feel most relaxed?

- When you're with close friends or family?
- While you're watching a film/TV show or reading a book?
- How about when you're taking a stroll?
- Out in nature?

- Or stroking a pet?
- In the bath?
- What is your breathing like in those situations?

Now give some thought to how you can allocate some time every day for relaxation.

NODDING OFF—BREATHING FOR SLEEP

When I was touring as a DJ, my sleep was all over the place and my body clock was perpetually confused by my sleep patterns. There were late nights, different time zones and way too much booze and partying. I used to "sleep in transit," which literally meant grabbing an hour or so on flights and coaches. Abnormal sleep was my normal. This is an extreme example. But many people struggle with sleep. Remember Ian, who, despite ending his day exhausted, spent his nights staring at the ceiling? Maybe you've experienced this yourself.

Up to 30 percent of people in developed countries suffer from chronic insomnia.[29] If you've experienced stress or anxiety, it's highly likely you've experienced trouble with sleep too. They go hand in hand. Think about it: our stress response is our fight-or-flight system, which gets you out of danger. If your brain is working on the basis that there might be a grizzly bear under the bed, do you think you're going to nod off easily? Absolutely not.

You need to calm your body and mind to allow it to activate your parasympathetic system and move into your relax response, and you can use your breath to do so. But there are some other helpful tips for nodding off that I'm going to share with you here.

SLEEP PREP

Get some natural light during the day

Sleep is closely connected to light. In fact, light is the principal control of our day—night cycle, influencing everything from body temperature and mood to metabolism and our level of the sleep hormone melatonin. A study in people with insomnia found daytime light exposure significantly improved sleep quality and duration. It also reduced the time it took people to fall asleep by 83 percent.[30]

Avoid blue light

There's a form of light you should avoid later in your day—the type emitted from your phone or computer screen. It's called blue light. Blue light fools the brain into thinking it's daytime, meaning that when you're on a device at night the body stops releasing melatonin, which then makes it hard for you to fall asleep.

So no phones or screens two hours before bed. If you *have* to use your phone or look at any other screen, then filter out the blue light. You can now do this on most smartphones by going into the settings. Some smartphones have a "night shift" feature that makes the color temperature of your screen warmer so it doesn't affect your sleep.

Develop a sleep routine

Be consistent with your waking and sleeping hours. This will create a regular sleep cycle that will tell your body when to feel tired. Optimize your bed environment—reduce light and noise, set the temperature to 59–66 Fahrenheit, and use a comfortable bed and pillows. Don't eat late (stop around three hours before your bedtime), and avoid foods that might interrupt your sleep, such as spicy or sugary food or foods that contain dairy. This may seem obvious, but watch your caffeine consumption. Caffeine stays elevated in your blood for six hours, so coffee after 3:00 p.m. should be a no-go! Watch your alcohol consumption too. It reduces night-time melatonin, so if you're having trouble sleeping it's best to avoid it.

THE CAFFEINE TRAP

Don't get me wrong, I love a brew. But if you need to improve your sleep or even maintain your natural energy levels, caffeine is not the best option. Although we think it gives us energy, what it actually does is prevent us from feeling tired, and this can play havoc on our sleep if we rely on it too much to fuel our day.

Remember from Chapter 1 how oxygen + glucose = adenosine triphosphate (ATP)? After ATP is used up it decomposes into adenosine. Throughout your day, adenosine builds up and binds with receptors in the brain, causing you to feel sleepy. It's like you're filling the adenosine bathtub all day, and when you go to sleep you pull the plug to empty the bath for the following day.

Sleep expert Matthew Walker explains that as the level of adenosine rises throughout the day, it needs some seats to sit on—these are the brain receptors.[31] Caffeine, however, blocks the adenosine receptors registering that you're tired. It's as if someone has taken away all the seats for the adenosine to sit on. Your body continues to produce adenosine, so when the caffeine eventually wears off, your brain is flooded with a massive dose of it and you "coffee crash." In this state, you struggle to get anything done, so you reach for another cup...and the cycle continues.

That surplus adenosine from excessive caffeine consumption is not always fully flushed from the body during a standard sleep cycle and can therefore contribute to the morning grogginess that many of us suffer from. This often encourages people to load up on more caffeine. It's a vicious cycle that results in even poorer quality of sleep—and tiredness all day. If you *have* to have your daily brew, limit it to one, and try not to have it after 3:00 p.m., so that you've got time for your body to flush out the adenosine when you sleep.

EXERCISE 15

ENTER THE LAND OF NOD: THE 4-7-8 BREATH

And now, for how breathing can help you to sleep. Welcome to the 4-7-8 breath. This is a sleep favorite, after celebrity doctor Andrew Weil boldly claimed that the method helps people get to sleep in one minute.[32] His evidence is limited to anecdotal reports from satisfied adherents, but what we do know is that it activates the parasympathetic nervous system, helping you wind down after a long day—the technique produces a very slight increase in carbon dioxide in your bloodstream when you hold your breath, which can make you feel drowsy. This isn't only fantastic for sleep prep, it's also a goodie to naturally tranquilize you back to sleep if you wake up and feel wide-awake during the night.

Here's how it's done. Before hitting the hay:

- Place the tip of your tongue behind your front teeth as if making an "LLL" sound.
- Close your mouth.
- Breathe in through your nose for a count of four, feeling your belly rise.
- Hold your breath for a count of seven. Keep calm and relaxed.
- Breathe out through pursed lips, keeping your tongue in

place, for a count of eight. Let your whole body relax and become heavy.

- Repeat for 10 rounds, or more as required.

Place the tip of your tongue behind your front teeth.

Close your mouth and breathe in for a count of four, then hold your breath for a count of seven.

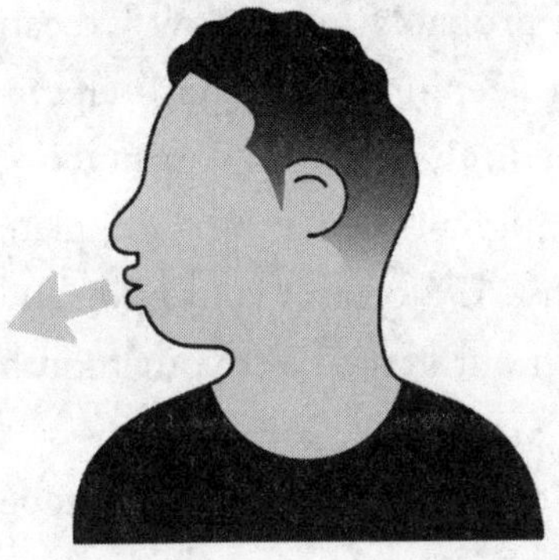

Breathe out through pursed lips for a count of eight, keeping your tongue in place.

DEEP SLEEP IS QUALITY

Do you wake up during the night? Do you snore? Do you wake up with a dry mouth? Do you suffer from sleep apnea? Do you wake up feeling tired? Do you wake up feeling stuffy? It's one thing falling asleep; getting *good-quality sleep* is quite another. And a good night's sleep is a big part of reducing stress during the day, as we've already mentioned.

To improve your quality of sleep, reduce your chance of snoring and wake up with a spring in your step, you should aim to improve your breathing when you're asleep. What's the quickest, most effective tool to help you do this? Shutting your mouth and breathing through your nose, of course.

It's vitally important that you breathe through your nose when you're sleeping. You can probably guess how we can guarantee nose breathing when sleeping. Yes, mouth tape.

If you can manage mouth tape in the day for 20 minutes or more, then you can graduate to night-time taping. Complete the 4-7-8 breath exercise (Exercise 15) to calm your body and mind and then place a piece of mouth tape over your mouth before you go to sleep.

If you consistently wake up in the morning with the tape still on, then you'll know you've retrained yourself to sleep while breathing through your nose. You'll notice this will make you breathe more through your nose in the day too. Win-win!

It's important that you don't do this after a night out boozing or if you have a congested or blocked nose. Only use one strip of micropore tape, and use your best judgment to keep yourself safe. If you feel uncertain or uncomfortable about doing this, simply stick to nose breathing as much as possible during the day instead; you should soon find that you'll pick it up naturally in your sleep as well.

It's vitally important that you breathe through your nose when you're sleeping.

SLEEP APNEA

Sleep apnea can cause people to suddenly stop breathing while they're asleep. When this happens, carbon dioxide accumulates in the blood, causing the sleeping person to wake up and gasp for breath. It's usually caused by mouth breathing, and the tongue falling back to obstruct the airway. This condition disrupts the sleep cycle and has a negative effect on health.

Sleep apnea is the most common disordered breathing while we sleep, with an estimated 1 billion people, between the ages of 30 and 69, affected globally.[33]

If this is you, don't worry. It can be usually resolved by building up to mouth taping at night.

IMPROVING DIGESTION THROUGH BREATHING

One of the biggest contributing factors to poor digestion and changes in appetite is stress. Of course, modern eating habits and processed foods are also to blame, but if you're a stressful breather—(chest, reverse, frozen, collapsed or grabber)—your body and brain will shut down your digestion system to prioritize functions that keep you safe.

In some people, chronic stressful breathing slows down digestion, causing bloating, pain and constipation, while in others it speeds it up, causing diarrhea and frequent trips to the loo. Some people lose their appetite completely.

Poor breathing that keeps you in a state of chronic stress can cause conditions like stomach ulcers and irritable bowel syndrome to develop. It can cause your weight to fluctuate from its healthy balance, causing either unhealthy weight loss or unhealthy weight gain.

EXERCISE 16

EASY AS PIE

To help your system digest food properly and remain in a good state of balance, you need to be in a parasympathetic state: the state of rest, *digest* and repair. The clue is in the description. So when you sit down to eat, before the fork even goes near the plate, you need to tell your brain and body that you're in a relaxed state so they direct blood flow toward the organs responsible for digestion.

- Before every meal, pause.
- Then, breathe in through your nose for a count of five, feeling your belly rise.
- Breathe out through your mouth for a count of 15.
- Repeat five rounds.

You may even wish to think about the amazing journey the food has been on to arrive on your plate. And say thanks for all the work and effort that's gone into its arrival on your fork. Gratitude promotes calm, and calm promotes digestion.

CUTS 'N' BRUISES

Activating your parasympathetic response is not limited to the times you want to feel calmer, sleep better or digest your food. Another powerful reason to activate it is to manage pain. And I should know. I was a rough-and-tumble kid. Throughout my childhood I had nicks, cuts, scrapes, grazes, sprains, fractures, breaks. This is what happens when you spend too much time with Tough Ted. But you don't have to be the next Evel Knievel (the motorbike stuntman who made the *Guinness Book of World Records* for suffering 433 bone fractures) to know how unpleasant it is to get hurt. You've no doubt stubbed a toe or burned your hand taking a dish out of the oven. Even if you haven't broken a bone, you've probably had one of those head-splitting migraines that makes looking into even the dimmest light feel like you're walking into the sun. Pain comes to us all, and in every shape and size—shooting, stabbing, burning, sharp, aching or throbbing. It can be short-term or long-term, it can be physical or emotional, and it can stay in one place or spread around the body.

Pain is simply your body's way of telling you something's up. Like stress, we all want to get rid of pain, but pain in some respects is a positive thing. It's another alarm bell. If you're injured or sick, it's there to tell you that you need to do something or stop doing something. For example, when you put your hand on a hot stove, your nerves send a signal to your brain and your brain sends you a message of pain. The pain screams at you to stop touching the stove and then take action to cool the skin. To use a more subtle example, if you walk on an injured ankle and it hurts, your body is telling you it's still injured and you need to stop walking on it so it can heal.

We memorize our pain responses, physical and emotional, to stop us from being in situations in which we feel this pain again. Try this now: close your eyes and imagine putting your hand on a hot stove. Did you feel some resistance to doing so? While memorizing pain is integral to keeping us safe from future pain, when pain becomes persistent or chronic it can also create stress and tension. So it's no wonder that many people with such pain try to disconnect themselves from their body—by drink, drugs or other means—in order to avoid it. The problem is that by doing this, they disconnect themselves from their body's innate ability to heal itself and find balance.

Many who feel even the slightest pain immediately reach for something to numb it. But when we try to avoid even low-level pain that's relatively easy to tolerate, we don't give ourselves a chance to adapt and grow. It's a bit like how, if you always drink coffee as soon as you wake up in the morning, you may never *actually* manage your energy levels. The same is true for drinking alcohol. If you reach for a glass of wine at the end of a hard day at work, you never let yourself learn how to unwind and switch from "work mode" to "rest mode." What's worse is that your body is very quick to adapt to the things that you put into it.

In the same way that faster breathing over a prolonged period of time can cause our body to adapt its biochemistry to remain balanced, our bodies are very good at changing their chemical baseline so that we need more and more of the things that we put into it to get the same effect, whether it's caffeine, booze or anything else. Even painkillers, though of course vital for certain ailments, can subtly do us harm. The good news is that we can learn to use our breathing to support us instead; it's a natural way to reduce and manage some low-level pain and the discomfort associated with it.

PAIN PERCEPTION

The perception of pain varies from person to person and there's no standardized test for its severity. It's subjective, influenced by your genetic makeup, your emotions, personality and lifestyle, along with your feelings, opinions and past experiences. If you're lucky not to have felt much pain in your life, your first experience of major pain, like a broken wrist or sprained ankle, could be a 10/10. Yet another person with a similar onset of new pain but with past experience of going into labor or having kidney stones might call this pain level a 5/10, compared with their past experience (which was a 10/10 for them). You also can get used to pain. When my grandma was a toddler she had osteomyelitis, a painful bone infection. This raised the bar of her pain threshold—in fact, her tolerance for pain was so high that as an adult she once broke her arm and didn't even realize it. Most of us would feel significant pain from that same injury. Because our pain is unique and influenced by our past experiences, and because our breathing maps these experiences, you can use your breath to help deal with it.

BREATHING TO PROCESS PAIN

The rhythm, rate and flow of your breathing also changes when you experience pain. This is because pain is stressful, so it stimulates the sympathetic nervous system. Stress hormones, including adrenaline and cortisol, are released and act as natural painkillers by binding to the opioid receptors in your brain to block the perception of pain. This is why a rugby player may not realize he has broken his nose mid-match, but experience agony afterward: pain is often delayed after an injury or accident because you're so pumped up on adrenaline at the time of injury. Regardless of whether your pain is chronic or dull, your sympathetic response gets activated and hormones released in this way.

One study has suggested that the way you breathe plays a big part in how you process pain.[34] This may come as no surprise to any mums who have experienced the natural pain relief that breathing provides during childbirth. Slow, deep, focused breathing is a powerful way to alleviate many forms of pain. For example, when you suffer acute pain, such as when you cut yourself chopping onions or bang your head on a low ceiling, your body reacts with shock, rising tension, an increasing heart rate and shorter, shallower breaths—a classic stress response. By slowing and calming your breath, increasing your out-breath length and using techniques like the ones we've practiced in this chapter, we can trigger our parasympathetic response, slow our heart rate, calm our minds, and allow tension to leave the body and the body to repair. Such techniques also reduce inflammation and the immediate pain response.

When pain continues for longer than 12 weeks despite medication or treatment, we call it *chronic*. Common examples of chronic physical pain include headaches, arthritis or back pain. There's a real need to find ways of coping with chronic pain that do not involve pharmaceutical drugs, which are often addictive and, in the long term, can potentially cause more problems than they solve. Harvard Health Publishing puts "deep breathing" at the top of its list of ways to control chronic pain, adding that it's "central to all the techniques" it lists for managing pain.[35]

Many types of chronic pain—including headaches, neck pain and back pain—can sometimes be caused by poor breathing in the first place. If you're overusing your chest and neck muscles instead of letting your diaphragm take the load of your breathing, they can get exhausted and tender. Practice "If in Doubt, Breathe It Out" (Exercise 14) for five minutes a day to help mitigate pain responses, even with chronic pain. If you're looking for more ways to reduce your pain, there are a few other tools that are good to add to your arsenal.

The way you breathe plays a big part in how you process pain.

EXERCISE 17

VISUALIZATION FOR PAIN RELIEF

Visualization is when we use our imagination to create an image or scene in our mind. Athletes often report that using it to supplement their training, or to imagine a big event going well before it happens, drastically improves their performance. Visualizing positive, pleasant images or scenes in our mind can distract us from pain and give us a greater sense of comfort and control. The more often you're able to redirect your focus away from pain, the weaker the neural pathways associated with it become. The more we focus on the feeling of pain, the stronger those pathways become. Visualization is therefore a helpful exercise when breathing through pain. And while some visualizations involve imagining yourself in a peaceful place, such as the beach or the forest, and are a great way to help you relax, for pain I like to use a more active visualization to help release it: picturing light.

Let me explain.

Let's apply this visualization to our "If in Doubt, Breathe It Out" technique (in for four, hold for four, out for eight). This can be an especially helpful exercise for low-level, temporary pain, such as muscle aches, injuries or menstrual cramps.

- Find a comfortable position, seated or lying down.
- Close your eyes and stay still for a minute to let your mind, body and breath settle.

- Now imagine a bright white light shining down on you, a healing light, a nurturing light.
- Just picture yourself bathing in this light; imagine it purifying your body.
- Breathe in for four, imagine breathing in white healing light.
- Hold for four, sending this light down your legs to your toes, your arms, hands and fingers, filling your whole body.
- Breathe out through your mouth for eight, letting your whole body really relax, and imagine the light leaving your body.
- Breathe in for four, imagining breathing in white healing light.
- Hold for four, this time directing the light to the area of pain, imagining the light dissolving your pain.
- Breathe out through your mouth for eight, imagining the light leaving your body along with your pain. Really try to visualize the pain leaving your body.
- Practice these last three steps for five minutes or more.
- If you experience chronic pain, do this exercise daily.

EASE YOUR POUNDING HEAD

Headaches are the worst. They can be triggered by all sorts: colds or flu, stress, eyesight problems, bad posture, lack of sleep, dehydration, too much booze the night before. Whatever it is that's causing your head to throb, here are three ways you can help reduce some of the tension they cause.

EXERCISE 18

PRESSURE COOKER: THE 7-11 BREATH

Sometimes when your head is pounding, the short, four-second breath hold of our "If in Doubt, Breathe It Out" technique can create a slight increase in pressure in the body. So with headaches, you may find the 7-11 breath more helpful. It's a very slow, relaxing breath, with a long exhalation to promote a parasympathetic response. This helps to calm the mind, relax the body and soothe that pounding head. It's a super-simple one; the title says it all.

- Find a comfortable spot, breathe in through your nose for a count of seven, feel your breath expand deep into your torso.
- Breathe out through your mouth for a count of 11.
- As you breathe out, make a conscious effort to relax your shoulders, jaw, face, forehead, even behind your eyes; let go of any tension.
- Repeat as required.
- You can add the light visualization to this breath practice too if you like, to help imagine the pain leaving your body.

EXERCISE 19

BEAT THE HEAT

We all get hot and bothered sometimes, and this can trigger the onset of a headache—too much time in the sun, hot flushes, or even the heat and tension that accompany anger or stress. Alongside making sure you rehydrate in these situations, this "Beat the Heat" technique is perfect to help cool and soothe the mind.

- Very gently clench your teeth, ensuring that the top and bottom rows touch, and open your lips slightly.
- Breathe in through your teeth using your diaphragm for a count of four, letting the cool air rush over your teeth and tongue.
- Breathe out through your nose for a count of eight.
- Aim for four rounds and see how you feel.
- You can also incorporate a little head movement into this technique, by lifting your chin to the sky as you breathe in and bringing it back down when you breathe out; slowly moving your neck can help blood flow and reduce the symptoms of a headache.

Breathe in through gently clenched teeth for a count of four.

Breathe out through your nose for a count of eight.

EXERCISE 20
TENSE TEMPLES

This "lion's breath" exercise stretches your temporalis muscles, which lie at the temples and attach to your jaw joints on either side of your head. These muscles often contribute to tension headaches. This exercise helps these muscles to release and relax as you breathe. It's also another goodie to help alleviate stress and eliminate toxins from the body.

- Find a comfortable seated position.
- Lean forward slightly and place your hands on your knees.
- Breathe in through your nose, using your diaphragm.
- Open your mouth wide, stick out your tongue and stretch it down as far as you can toward your chin.
- Breathe out forcefully, making a "Ha" sound as you empty all the air from your lungs.
- Breathe normally for a few cycles.
- Repeat four times.
- Finish by breathing slowly, deeply and gently for one to three minutes.

Breathe in through your nose while sitting down with your hands placed on your knees.

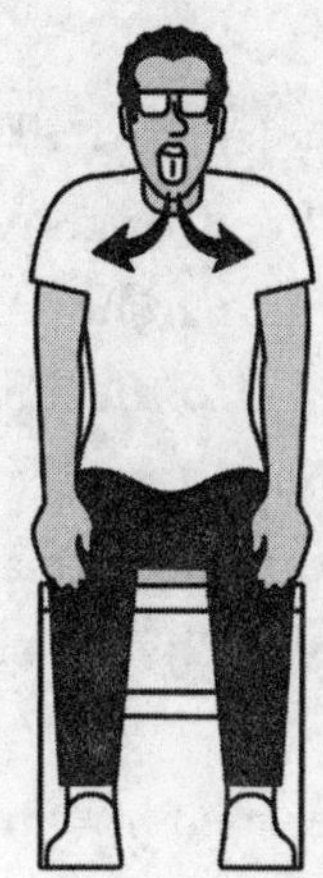

Breathe out through your mouth with your tongue stretched down to your chin.

★★★

You've now learned all about the basics of breathwork, and what your breathing says about you by learning more about your breathing archetypes. You've learned how breathing is closely related to the autonomic nervous system—your sympathetic (*S* for "stress") and parasympathetic (*P* for "peace")—and how you can take advantage of this to move into a state of calm, to get better sleep, to improve your digestion and to process pain.

You've learned how to start fixing your breathing, by using your nose and diaphragm, slowing your flow, improving your carbon dioxide tolerance, and are now experiencing more balance throughout the day. You've also learned the "If in Doubt, Breathe It Out" technique, among a few other handy exercises, that you can turn to whenever things feel out of control.

Going forward, you should continue to practice the following exercises daily. In addition, breath awareness throughout your day is key. Really try to pay attention to your breathing and how it changes. Keep testing your carbon dioxide tolerance (Exercise 12) every week, and refer back to the intentions you set at the start of the book to set your sights on where you're going.

YOUR DAILY BREATHWORK PRACTICES:

Exercise 5: 'Tape It Shut.' Build up to twenty minutes daily, then switch to taping when you are sleeping.

Exercise 8: Release Diaphragm Tension

Exercise 9: Push the Sky

Exercise 10: Diaphragmatic Breath

Exercise 11: Magic Ratio Breathing

Exercise 13: Slowly Does It

Exercise 15: Enter the Land of Nod: The 4-7-8 Breath (before bed)

Exercise 16: Easy as Pie (before eating)

WHENEVER YOU NEED A LITTLE EXTRA HELP:

Exercise 4: Nose-Unblocking Technique

Exercise 14: If in Doubt, Breathe It Out

Exercise 17: Visualization for Pain Relief

Exercise 18: Pressure Cooker: The 7-11 Breath

Exercise 19: Beat the Heat

Exercise 20: Tense Temples

While these exercises will help you to breathe better, sleep soundly and manage your stress and pain, the true power of breathing on your health and well-being reveals itself when we do some deeper work. Over the next three chapters, we're going to be going on a journey further into your body and mind. We'll be exploring how your breathing is affected by your emotions and trauma, and how this, in turn, can affect your health and well-being. Before you say, "I've never experienced anything traumatic. This next section isn't for me," I want to tell you that our body and mind store all sorts of experiences, big and small, as trauma. This shows up in your beliefs, your breathing, your behavior and your overall physical, mental and emotional health. In this next section, I'm going to teach you how to begin to stabilize your emotions, unpack your traumas, understand your beliefs, and create long-lasting positive change to your breathing and your life.

PART 2:
DEEPER WORK

5
UNDERSTAND YOUR EMOTIONS

A WORLD OF EMOTION

It's your birthday. You've been trying to organize a get-together with your mates to celebrate. Jonny—well, he said he's away this weekend, René's got the final part of her coursework to complete, and Leon and Inés don't even reply to the group message. Even your family say they're busy, but they'll try to do something later in the week. You feel a bit bummed out, to say the least, and a heaviness weighs on you. Your breathing is frozen. Then, finally, you get a message back from Chris.

"I can meet you after work," she says.

You perk up a little. "Let's meet at the bridge on the canal and we can walk over to the café for a bite to eat. I'll book a table for us."

You're still a little deflated, but excited that at least one person wants to hang out. So you go to meet up with Chris and head to the café.

"Hello," says a grinning waiter. "Your table is at the back. Just head through the double doors." You pass a large group, and you feel crap that this wasn't you today. But as you push the doors open, kapow, streamers fill the air!

"Surprise!" a chorus erupts. "Happy birthday to you, happy birthday…"

You scan the room. All your favorite people in the world are there: Jonny, René, Leon, Inés, all your family, the cousins you never see, even your best mate who's moved to the other side of the world has traveled back to celebrate with you. Endorphins fill your body, your breathing flows naturally and you're overwhelmed with joy. You burst into tears.

Life would be pretty dull without emotions. They're the spice of life. Within an instant things can shift and change, rejection can become joy. Emotion is like music—it flows, it's felt and it's hard to describe with words.

EXPRESS YOURSELF

Emotions, as we all know, are complex things. Some, like grief, can be made up of a number of "simpler" emotions, such as sadness, anger and frustration. Others can persist over time and transform into different emotions, such as unexpressed anger becoming resentment. Emotions simply don't fit neatly into boxes. But the important thing to note is that the kinds of emotions you're likely to experience on a daily basis—simple emotions such as fear, stress, happiness or anger—*can* be addressed and processed, and this will give you more resilience, emotional intelligence and control in your daily life.

The word *emotion* comes from the Latin *emotere*, which means "energy in motion." When you feel an emotion, you're simply recognizing certain energies moving through your body, a *felt charge* based on what you're experiencing. Emotions may cause you to scream and shout after receiving good news, collapse on the floor sobbing when you can't find your house keys or to withdraw completely after an argument.

Emotions are felt responses to situations and play a key part in your daily life. You probably notice yourself experiencing a wide range of emotions, and this is because your brain and body are constantly communicating to keep the chemistry in your body balanced. It's an example of homeostasis, by which living organisms try to maintain relative stability despite changing circumstances.

When there's a change to your body's chemistry, such as

when you feel an emotion, your body attempts to regain balance by releasing that emotion—what we call *emotional processing.* When you laugh, cry or shout, the body is trying to release and let you process energy. Breathing is an important part of this; it's the mechanism that helps facilitate the release of emotion. Every emotion comes with a different breathing pattern and flow.

Try this.

Mimic the action of laughing right now. Can you feel your breathing rhythm change? Your out-breath jitters as you expel stale air from your lungs. Laughing expands the tiny air sacs in the lungs, and creates more room for fresh oxygen to enter. OK, now let's try to mimic crying. How is your breathing flowing now? Is it more restricted, short and shallow? OK, one more: I want you to mimic rage. What happens to your breathing? What happens in your body? Can you feel it tense up and your muscles contract?

Positive emotions are felt as an expansion of the body and breath, and negative emotions as a contraction. Sometimes we also consciously restrict our breath as a means of controlling an emotional outburst, positive or negative—like that time you had to hold your breath to stop yourself laughing at your school teacher, or when you held your breath to keep the tears from flowing in front of colleagues at work.

Emotions, both positive and negative, lead to behavioral responses, where you either *express*—your breathing and emotion flow; you laugh, you cry, you shout—or you *repress/suppress*, which often correlates with your breathing being restricted and your emotions becoming "trapped"—they remain in the body, unexpressed and therefore unresolved.

When your emotions are out of balance, or you do not allow yourself to express them, it causes problems. You might have an emotional outburst and slam some doors, you might say something you regret or you might try to squash the feeling completely by avoidance, substance use or distraction. But sweeping emotions under the rug does not help, and has been shown to cause stress, anxiety, sleep issues, depression, addictions and, over time, even disease. This is why it's super-important to express yourself (in a safe and appropriate way) and allow emotional processing to happen as it's supposed to. We often call this processing *integration.*

After much experience in my practice, I'd say that most bad breathing habits and the corresponding archetypes they form stem from holding patterns, when people try to sweep their emotions under the rug. It's important not to avoid your emotions but to understand them and get better at feeling them, and to free your breathing from trapping them. When you're in tune with your emotions, you make better decisions and have more control over your actions. Your breath flows freely, and you feel healthier and happier. In this chapter we're going to learn how to work with your emotions, not against them.

LET TEARS FLOW

Although crying disrupts the rhythm of our breathing, it's nonetheless an essential action and a common way we express emotion, both positive and negative. One 2014 study found that crying had a soothing effect as it activates the parasympathetic nervous system to reduce the stress of emotion and releases a feel-good chemical called oxytocin, which relieves physical and emotional pain, and lifts mood.[36]

When humans cry in response to stress, it has also been shown that their tears contain an increased number of stress hormones, which researchers hypothesize could help reduce stress-hormone chemical levels in the body during an emotional experience.[37]

Another study has shown that crying is linked to attachment behavior and rallies support around us, a preverbal tool we learned as a baby before we could communicate through words.[38]

Every emotion comes with a different breathing pattern and flow.

NINETY-SECOND RULE

Have you ever noticed how young children are pretty free with their emotions? One moment they're having a tantrum, the next they're happily playing like nothing has happened. One moment they're warm and loving, the next they're huffing, prickly and unkind. This is because they're forming a personality and haven't yet developed clear beliefs about the world. Although it can be stressful to bounce between emotions too quickly, children generally let their emotions move through their body as emotions should. And while it's not always appropriate to express our emotions in the way a child does, it's important to recognize that if we suppress our emotions, the experience that caused it ends up as another brick in our bag.

Harvard brain scientist Dr. Jill Bolte Taylor, author of the bestselling book *My Stroke of Insight: A Brain Scientist's Personal Journey*, found that a simple emotional reaction, as expressed in the mind, and its corresponding chemical process within the body, only lasts 90 seconds. If you fully recognize and feel your emotions—let's say you notice and wholly allow a surge of anger to flow through you—it dissipates within 90 seconds.

This blows my mind. Sometimes I've felt my emotions last days, weeks, years! If you find that an emotion is remaining after that 90-second window, then it's either because you've tightened up your body and held your breath to trap it, or you've created a story around that emotion. This could happen consciously or unconsciously. For example, when we begin to feel anxious, the story some people tell themselves might sound something like this: "I feel anxious about the future because I can't do anything and I'm worthless." Through this anxious thought, your breathing becomes anxious, the emotion intensifies, potentially spirals and lasts longer. And the more we replay the memories of the experiences that created that emotion in the first place and the more we obsess about the corresponding thoughts we've attached to it, the more we remain stuck in an emotional cycle, and its corresponding breathing pattern, that gets increasingly difficult to break free from.

Of course, when it comes to emotional processing, there's a big difference between the feelings of disappointment you get when the shop has run out of your favorite ice cream and the heartache you feel after a breakup. This relates to the difference between simple and complex emotions that I described earlier. Certain experiences, such as heartbreak, can be so devastating that they produce a complex emotion right away. In other situations, simple emotions like disappointment can end up developing into a complex emotion, such as deep regret, over a period of time. You may recall this happening in the last chapter, in the "before the job interview" example.

Complex emotions are much more difficult to process, but many of the complex emotions that we feel begin as simple emotions that have spiraled out of control. So first of all, let's tackle this with a method of processing our "simple" emotions—the ones we all have to confront on a daily basis—to stop them from intensifying and spiraling into more complex emotions.

EMOTIONAL SPIRALS

Fortunately, emotional spiraling can work both ways, positive and negative.

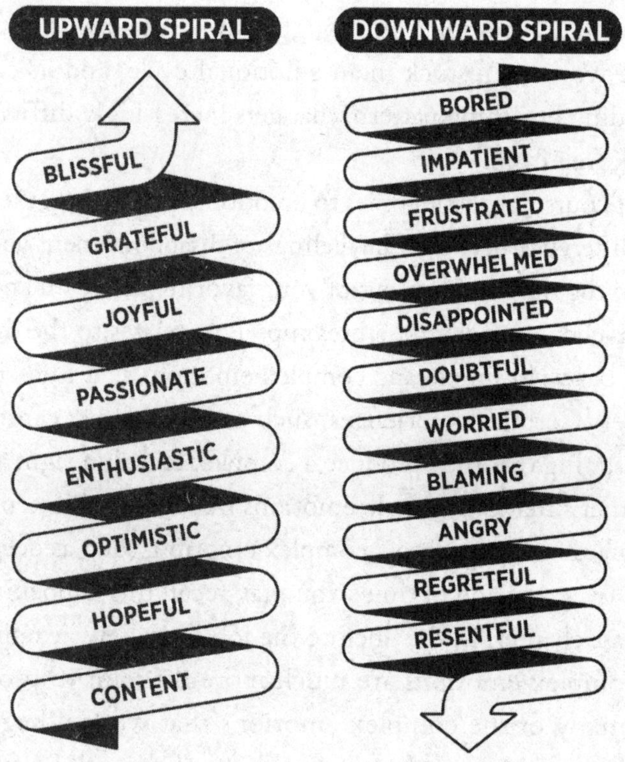

A powerful and safe way of breaking free of a negative emotional spiral that also allows you to fully feel the emotion currently moving through your body is the "Recognize–Breathe–Reframe" technique.

EXERCISE 21
RECOGNIZE—BREATHE—REFRAME

This technique involves you becoming more aware of your feelings and thoughts as they appear throughout the day. It encourages you to accept them, process them and investigate them over the course of 90 seconds, before letting them go.

RECOGNIZE

The first step is to recognize your emotion.

- What are you feeling?
- Notice it in your body and become familiar with how it feels.
- Where do you feel it in the body? Be specific. Is it in your chest or your abdomen? Does the feeling stretch elsewhere?
- When you recognize emotions, you can start to consider your response, instead of having a knee-jerk reaction to the situation.
- Now label the emotion that you're feeling. Acknowledge it and accept it—but don't become it.
- For example: "I'm angry and frustrated" becomes "I'm feeling or experiencing anger and frustration."

BREATHE

Now that you recognize your emotion, we can begin to breathe through it. Breathing through your emotion will allow your body to begin to process and integrate it.

You may wish to let yourself just breathe calmly for at least 90 seconds; your body may intuitively pick a rhythm that feels best to move that emotion on. If that sounds too advanced or abstract, read on.

I've found the best practice to help shift an emotional state is something that I call a "humming breath." It helps free the mind of agitation, frustration, anxiety and anger. As this practice involves some sound and movement, you may wish to find a place where you feel comfortable, especially if you're in a public space.

- Breathe in through your nose for a count of five, feeling your belly rise.
- Make a humming sound for as long as you can as you shake out your hands, arms and body. Imagine shaking off the emotion as you make the sound.
- Repeat for four rounds or more (to take you over 90 seconds).

REFRAME

So you've recognized your emotions and breathed through them. Your final step is now to reframe them. This means to think about them in a different way.

You reframe your emotions by asking yourself four questions:

1. What happened to make me feel this way?
2. Is there an explanation that makes sense?
3. What do I want to do right now? (This is an opportunity to acknowledge a desire to feel a different way emotionally.)
4. Is there a better way?

So let's say a job interview didn't go your way. The sense of disappointment, combined with the weight of expectation, could ordinarily cause you to enter a negative emotional spiral. The "Recognize–Breathe–Reframe" technique could instead help you process this emotion, beginning a spiral of positive feeling upward. And remember, due to our negative bias, if an emotion is not spiraling upward, even slowly, it's spiraling downward.

Let me show you:

Q: What happened to make me feel this way?

A: I was rejected for my dream job.

Q: Is there an explanation?

A: Yes, I don't have enough experience.

Q: What do I want to do right now? (Acknowledge your emotional desire.)

A: I want to throw my phone against the wall.

Q: Is there a better way?

A: Yes, ask for feedback and build more experience so that I can apply for something similar in the future.

It may take a bit of practice before this becomes a habit, and your responses may not always run as smoothly as the example above. But with practice, going through these steps in your head will become easier and more effective. Managing your emotions in this way, with logic, reason and compassion, will also expand your capacity for productivity and self-care, promoting healthy relationships with yourself and others.

When we have the awareness that simple emotions last for 90 seconds and that if we let them flow through us without resistance, we can let them go, we can start to reflect on our emotional triggers and reactions in a deliberate and rational way.

The challenge arises when we have repeated experiences that create the same emotions, or when we're stuck in a negative emotional spiral. In these situations, the body can get so used to the chemical cocktail caused by these emotions that it can become addictive, like a drug.

TOXIC POSITIVITY

Some emotions are deemed more acceptable than others. As someone who always tried to be a positive person, I spent much of my life focusing on the good in situations and not the bad. While this is helpful sometimes, it is important to recognize that negative emotions are a part of life; they never simply disappear. I was sweeping them under the rug, instead of allowing myself to express them.

The practice of focusing on the good in every situation has also given rise to the idea of "toxic positivity," an obsession with positive thinking and suppressing negative emotions. This way of thinking—putting a positive spin even on events and experiences that are profoundly tragic—risks silencing negative emotions, belittling the experience of loss and bad times, and putting pressure on people to be happy even when they're not. Ultimately, it can do more harm than good. It can even turn your simple emotions into complex ones. Instead, express your feelings and thoughts, but don't identify with them.

THE MIND–BODY CONNECTION

This relationship between our thoughts, feelings and our bodies is called the "mind–body connection." Neuroscientist and pharmacologist Candace Pert was a prolific contributor to this field, publishing more than 250 research pieces on the topic. She discovered that thoughts and feelings trigger chemical changes in the body through the release of tiny chemical proteins called neuropeptides (NP).[39]

Your cells produce hundreds of different neuropeptides, each with its own function. Every thought triggers a different type. Emotions such as joy and gratitude release NPs such as feel-good hormones, including endorphins or oxytocin. Emotions like stress, fear or anger will elicit NPs such as cortisol and adrenaline, which are helpful when an immediate response or action is required, but become harmful when released over a prolonged period of time and can weaken the body. Any prolonged mental state that produces negative emotions such as fear, anger, worry, guilt and shame will inevitably change the chemistry of the body, and can potentially cause an addiction to the chemical cocktail they produce.

A CHANGE OF HEART

Researchers from the HeartMath Institute based in Boulder Creek, California, recorded individual heart rhythm patterns as subjects experienced different emotions. Incoherent heart rhythm patterns, when the heart beats in an erratic, disordered way, were characterized by irregular, jagged waveforms, and were associated with times of stress and negative emotions such as anger, frustration and anxiety. Coherent heart rhythm patterns, when the heart beats in an ordered way, were characterized by regular, smooth, continuous waves, and were typically observed when an individual was experiencing a sustained positive emotion, such as appreciation, love, compassion or flow.

EMOTIONAL ADDICTION

I was booked to play a DJ gig in Mexico. It was my first one in a while, as I had taken time out after Tiff died and dived deep into the world of breathwork. Yet despite all the "inner work" I was doing, the moment the rush to catch the flight began, I transformed, like the Hulk, into the worst version of myself: Airport Stu.

Airport Stu had shown up throughout my life whenever I was running late. But this time, I was aware of the unconscious patterns in my thoughts, actions and breathing that were hijacking my mind. I started to understand how I became Airport Stu.

But that awareness alone wasn't enough to stop Airport Stu from being Airport Stu. And just as some people get road rage, Airport Stu gets airport rage. It began in the queue for security, as I checked my phone's clock impatiently, silently seething at everyone in front of me.

Up ahead, an older lady kept forgetting that things were in her pockets as she went through the body scanner, *bloody amateur.* Some people might have thought she looked sweet, but oh no, not Airport Stu. She was going to make me miss my flight. And what *was* that guy doing? You take your toiletries *out* and put your bag *on* the fucking conveyor belt. For fuck's sake, it's not that hard. Why are they all so slow? They better not make me miss my flight.

Now it was my turn. Liquids out, belt off, bag on. I cruised through. I'm slick, I'm seamless. Airport Stu has it down, it's everyone else's fault if he's late. I made it through security with five minutes to spare, raced through the assault course of perfume-spraying duty-free shoppers, leaped over kids wheeling animal cases outside Boots Pharmacy, and was gearing up for the final sprint when I caught sight of the announcement screen. *Flight delayed. Wait at gate.*

Feel-good chemicals 239

my body and sent a wave of smug flowing through me. I'd made it. Heart still pounding, I bought a coffee, wiped the sweat from my forehead, grabbed a seat and took a couple of deep breaths. I then opened my book, *Why Zebras Don't Get Ulcers* by Robert M. Sapolsky, and downed a large gulp of my triple-shot latte. The older lady strolled past and smiled. A thought floated through my mind. Why did I put myself through all that? And by chance, the book answered. The body becomes addicted to stress in the same way it gets addicted to drugs. At first you need a little, and then the more you get used to it, the more you need.

The penny dropped. *I was a stress junkie.*

The body becomes addicted to stress in the same way it gets addicted to drugs.

EXERCISE 22

ARE YOU ADDICTED TO STRESS?

Have a look at the below questions and answer them as honestly as you can. If you answer yes to more than half of them, you might just have your own version of Airport Stu…

- Do you love a tight deadline?
- Do you leave things until the very last minute?
- Do you have a difficult time doing nothing?
- Do you think about work while lying on the beach?
- Do you get FOMO?
- Do you check your phone while you're watching TV?
- Do you feel as though there's never enough time to get things done?
- Do you have a to-do list as long as your arm?
- Do you ever feel as though what you've achieved at work for the day is not enough?
- Do you feel as though you're constantly running from one thing to the next?

When we think of addiction, we usually think about compulsively taking an external substance like drugs or alcohol, or we think of addiction as a form of pleasure, like sex. But it's also possible to be addicted to the chemical cocktails created when we repeatedly feel emotions—even if these emotions make us feel miserable.

As we've explored, the brain provokes chemical reactions in the body in response to certain emotions. An emotional addiction begins when your body becomes dependent on a particular chemical response. If you repeatedly feel an emotion, or if an emotion is trapped in your body, the body establishes a new chemical baseline. This in turn changes the reward center of your brain. When this happens, an emotional addiction develops. You become reliant on a specific chemical reaction associated with the emotion, and your unconscious mind may seek it out just to feel "normal," even if it's something you don't consciously wish to feel.

For example, if anger is the emotion you feel most often, you may become addicted to this state and you might find yourself getting angry whenever you're uncertain about something. You may even feel a sense of relief as this emotion rises within you because the rush of that particular emotion triggers the reward system in your brain. Emotional addiction is harder to spot than substance addictions because it happens in our unconscious mind, and we often dismiss it as simply a personality trait.

I'M *SOOO* BUSY

Modern Western society encourages our stress addictions. How many times have you heard your mates boast, "I'm so busy," followed by "How was your day? Crazy busy?" And how often have you felt at the end of the day that you haven't done enough? In our minds, busyness seems to equate with importance, even though being busy doesn't necessarily mean being productive.

This need to be busy all the time is yet another addiction. Our smart devices ensure that we can keep ourselves on the busy treadmill even during dinners, holidays and social gatherings. We're always on, glued to our busy life, with adrenaline, cortisol and dopamine drip, drip, dripping in. Money can also play a role in this cycle, as self-worth is seemingly measured in dollar signs. Society tells people that they're worthwhile if they make more money, which causes people to overwork and keep busy.

So the B-word is now out of bounds. It's now a swear word. If you say it, you've got to put a dollar in a jar. What you need to do is reframe the B-word, and challenge yourself to be honest and say what you've actually been doing. Like, "I'm working on an exciting project" or "I've been chatting to my mates online." Whatever you're doing, don't just say, "I'm busy." This will help you to identify what you're spending your precious hours on, giving some focus, positive energy and motivation toward the tasks or projects that matter to you.

CARLO

When Carlo first came into my studio, he started talking and didn't stop until he ran out of breath, about 15 minutes in. He had all the signs of a breath grabber/breath controller hybrid, someone with something to prove who was also trying to take control of things in his life. He'd come to me because he often found himself in situations that made him feel an emotion he didn't like, an emotion that no one likes: guilt. This emotion arose in a whole variety of circumstances—because of his food choices, something that happened at work, his relationships, what he'd done, what he hadn't done, what he could have, would have or should have done. Even the way he spoke to himself could give him this feeling. He habitually blamed himself and felt guilty for things that were either justifiable or not his fault. He felt stuck in a loop, repeatedly experiencing situations that created that same feeling of guilt.

Carlo was feeling this emotion on repeat as his body had set up a new chemical baseline, so much so that both his body and unconscious mind began to think that this guilt state was normal. He was addicted to being in this state, and as weird as it might sound, whenever he experienced guilt, his brain's reward system and the cells in his body got a hit from it. If Carlo hadn't received his "required" dose in a while, his cells would send a signal to his brain for him to begin ruminating on a past situation that made him feel guilty; sometimes he even unconsciously sought out experiences that made him experience more of that emotion. And thus the pattern continued. This kind of addiction is incredibly common, and it can happen with all sorts of emotions. You can imagine how destructive this can be. Carlo required a number of sessions to release and reset both the breathing pattern he was stuck in and to rediscover his self-worth.

If feeling emotions in this way sounds familiar, remember that awareness is the first step in making your unconscious patterns conscious and starting to effect change. You may not initially think you have an emotional addiction because it's become so deeply woven into you. It feels like "part" of you. So like I told Carlo, take time to be aware of how you feel in daily situations—at work, when you're with friends or family, and even look for signs when you're scrolling on your social feeds. Look for any patterns of feeling a certain emotion and notice any changes in your body or breath, where you hold any tension. This may give you clues as to precisely which emotion has got its grip on you. We'll be exploring ways of working through emotional addictions later in this chapter.

WHEN EMOTIONS DON'T FLOW

Airport Stu didn't begin with a big dose of gamma radiation, like the Hulk did. There are two ways you can become addicted to an emotional state. One is where a certain emotion becomes deeply entangled with your beliefs about yourself (we'll be exploring how to overcome this in the next chapter). The second is when an emotion gets trapped. This can happen when you don't allow yourself to fully feel and process your emotions for 90 seconds.

When your emotions don't flow naturally, they can cause your chemical baseline to change. This works in exactly the same way that your chemical baseline changes when you're addicted to a certain emotion. This trapping can happen in one of two ways. First, unconsciously, by repression, which involves your brain blocking unwanted emotions because they contradict your beliefs. Your unconscious mind might try to protect you from feeling pain or embarrassment, because you've been taught to believe showing "weakness" is wrong. Second, consciously, by suppression, whereby you deliberately try not to feel emotions that you feel are inappropriate. Like not feeling it appropriate to cry in front of your work colleagues, or laugh when you "shouldn't." One of the most common ways in which you control emotions and prevent them from being experienced is holding your breath and creating tension in your body.

Tension, even if it's just the smallest micro-contraction,

stops the natural processing of emotion. It freezes your emotion in time, completely halting its energy in motion. And by restricting the natural open flow of your breath when you're feeling an emotion, you make sure it stays in your body. Some traditions believe that the trapped emotion—whether you've held back joy, laughter, anger or something else—can manifest as physical pain. Chinese medicine, shamans, Reiki experts and somatic therapists link areas of pain and tension in the body to certain thoughts, feelings and behaviors. Chinese medicine, for example, one of the oldest healing practices in the world, going back thousands of years, links the heart to a lack of vitality, the lungs to grief and sadness, and the liver to anger and frustration. Such associations have also started to make their way into modern science, with researchers finding not just a link between an emotion and a physical pain, but between the source of an emotion and a pain. For instance, a combined research team from the University of Virginia and Columbia University found that something as specific as financial worry can cause lower back pain, neck pain and shoulder pain. According to lead study author Eileen Chou, results from six studies, one with data from 33,720 households, establish that economic insecurity produces physical pain, reduces pain tolerance and predicts over-the-counter painkiller consumption.[40]

By the time you begin to see physical symptoms such as those described above, you've probably been experiencing a specific mental state or trapped emotion for quite some time.

Of course, you might say that emotion is already physical. If we're sad, we frown, we hunch over, perhaps cry. If we're stressed or angry, we tense up, maybe shake our fists and furrow our brow. But the link is even closer. A Yale study suggested that stress and physical pain are in fact two sides of the same coin. Researchers found a "significant neuroanatomical and physiological overlap" between discomfort in the mind and discomfort in the body.[41]

While peer-reviewed studies into these mind–body connections remain few and far between, a link between emotional and physical pain is something I see in most of my clients. Research in this space is ongoing and there's certainly a growing appreciation that the body is a single interconnected unit, rather than something reducible to individual parts that have no relationship with one another. At the very least, I believe that emotions can be a helpful reference point when exploring the origin of a pain that has no obvious cause. In my experience of working to support people through pain, dealing with difficult or trapped emotions tangibly lessens the discomfort that the pain produces.

Consider these questions to explore if any pain you feel could be linked to emotions or trapped emotions. This short questionnaire will help you identify those associations. We'll work to process them in our next chapter.

- Do you have aches or pains that have seemingly come from nowhere?
- Where are they located in your body?
- Is the weight of the world on your shoulders, causing you shoulder pain?
- Are your financial concerns causing you back pain?
- Is your neck pain due to feeling unsupported in your life?

- Is there an ache or pain in your life that may be linked to an emotional experience?
- Have you suffered an accident or injury? What was happening in your life around that time? Can you recall your mental state and how you were feeling?

MAPPING TRAPPED EMOTION

I love maps, and in my treasure hunt for knowledge about the power of breathing to treat pain, trauma and trapped emotion in the body, I've come across a wide variety of them. Many are rooted in Eastern traditions, especially Chinese medicine and the Indian yogic tradition, but there are some from Western science too. A Finnish study of more than 700 participants published in the *Proceedings of the National Academy of Sciences* mapped the influence emotions have on our bodies in consistent ways.[42] Anger was felt most strongly in the head, and happy people felt warmth all the way through their bodies to their fingers and toes. People who felt depressed reported numbness, with little feeling in their head and torso, and almost nothing in their limbs.

There are differences of opinion over exactly where certain emotions or thoughts are mapped onto the body, but also significant agreement. The throat, for example, is widely understood to be where discomfort relating to expression originates, while the holding back of certain emotions, such as anger, is often housed in the jaw. The chest tends to be where grief and sadness are mapped, and the back is usually where guilt is held—I could immediately see tension in Carlo's back when he started breathing in my studio.

A body map, such as the one on page 232, is very helpful in showing where tension is originating in your breathing cycle and may help to identify what is causing your physical or emotional pain or disease. The integrative breathwork school Transformational Breath® uses body maps like these alongside positive affirmation to help release tension in the breathing cycle. This is something I have found to be extremely effective in sessions. If you suffer with pain throughout the day, try to notice where in the body this discomfort or tension presents itself. Pay attention to anywhere that's quick to freeze up or get sore, and note whether there's an area in your body that your breath doesn't seem to flow freely into. The areas in question could give you a clue as to what kind of emotions you've got trapped. I have also included some positive affirmation statements in the tables. These are short phrases you can repeat to help promote a release of tension in a specific area. You may even want to use these to update the affirmation statements you set at the beginning of the book.

BODY TENSION MAP – TRAPPED EMOTIONS

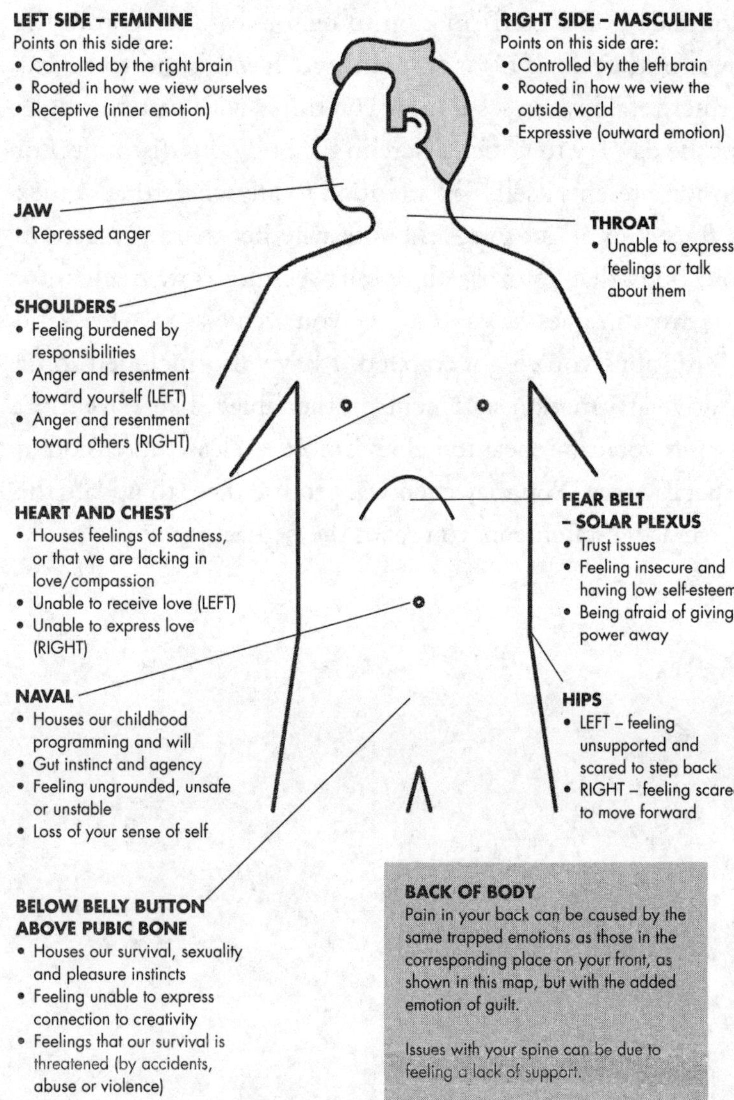

BACK OF BODY

Pain in your back can be caused by the same trapped emotions as those in the corresponding place on your front, as shown in this map, but with the added emotion of guilt.

Issues with your spine can be due to feeling a lack of support.

AFFIRMATION STATEMENTS TO RELEASE TENSION

<table>
<tr><th colspan="3">FRONT OF BODY</th></tr>
<tr><th></th><th>LEFT</th><th>RIGHT</th></tr>
<tr><td>Jaw</td><td>I acknowledge my anger without losing control.</td><td>I release all my anger.</td></tr>
<tr><td>Throat</td><td colspan="2">I let my voice be heard.
I speak my truth.
I express myself openly and fully.</td></tr>
<tr><td rowspan="2">Shoulders</td><td colspan="2">I am at peace with what is happening and what will happen.
I welcome mistakes as opportunities to grow.</td></tr>
<tr><td>I love and forgive myself.</td><td>I love and forgive everyone.</td></tr>
<tr><td rowspan="2">Chest and heart area</td><td colspan="2">I am grateful.
I fully love and embrace who I am.
I give and receive love effortlessly and unconditionally.
I follow my heart.</td></tr>
<tr><td>I love and accept myself.
I show compassion for myself always.</td><td>I lead with love.
I feel compassion for others.</td></tr>
<tr><td>Solar plexus fear belt</td><td colspan="2">I trust.
I let go of control.
I am free of fear, tension and stress.</td></tr>
<tr><td>Naval lower abdomen</td><td colspan="2">I am here.
I am safe.
I am grounded.
I am powerful, rooted and strong.</td></tr>
<tr><td>Below belly button</td><td colspan="2">I choose life.
I am connected to my body.
I am grateful for my body.
I pay attention to my body's needs.
I am open to receive pleasure.
Creativity flows through me.
I am in perfect flow.</td></tr>
<tr><td>Hips</td><td>I step back, pause and gain perspective.</td><td>I move forward with joy and ease.</td></tr>
</table>

BACK OF BODY
I treat myself with respect and kindness.
I let go of my guilt and shame.
It's safe to be with me.
I am at peace with the past.
As I forgive, I grow stronger.
I release the past and allow myself to move forward.
I am supported and stable.
I am abundantly rewarded.
Self-love is my priority.
I am enough.
Opportunities are everywhere.
I choose to let go of the past.
Today, I choose to live in the moment.
I allow myself to move forward in life.
I am learning and growing every day.

A KINK IN THE HOSE

A central concept of Traditional Chinese Medicine (TCM) is that vital energy called "qi" moves along energy pathways and brings life into all living cells. When emotions become trapped, qi becomes trapped, and TCM believes this injures organs and opens the door to disease. The energy in your body is like water in a hose; a kink in the hose will stop its natural flow. If an emotion is trapped, it creates tension that can lead to health complications and restrictions on our breathing. A lot of our aches and pains, if TCM is right, could be the result of an emotional experience that we've not allowed ourselves to fully feel.

EVERYTHING IS VIBRATION

Some of the greatest scientific minds, including Albert Einstein and Max Planck, birthed a series of discoveries that launched the notion of quantum physics. A principle of quantum physics is that everything in the universe, seen and unseen, consists of tiny building blocks called atoms. These are the smallest units into which matter can be divided, and they vibrate at certain frequencies. Zoom in close enough to anything and you will find atoms packed together at different densities. The seat you're sitting on, the clothes you're wearing, the music you're listening to. Even the air we breathe is made of atoms.

Everything is made of these teeny-weeny bits of vibrating energy. Even your thoughts and feelings have their own vibrational frequency. Candace Pert took this further, saying: "We're not just little hunks of meat. We're vibrating like a tuning fork—we send out a vibration to other people. We broadcast and receive."[43] This is sometimes called the law of vibration. I've treated thousands of people to great effect on the basis that emotions like anger, anxiety and guilt resonate at very low frequencies, while contentment, optimism and gratitude resonate at the highest. My treatments revolve around creating a high-frequency environment in the body, through which we can begin to transform those negative feelings into positive ones. And the reason behind why we can do this is all down to something called entrainment…

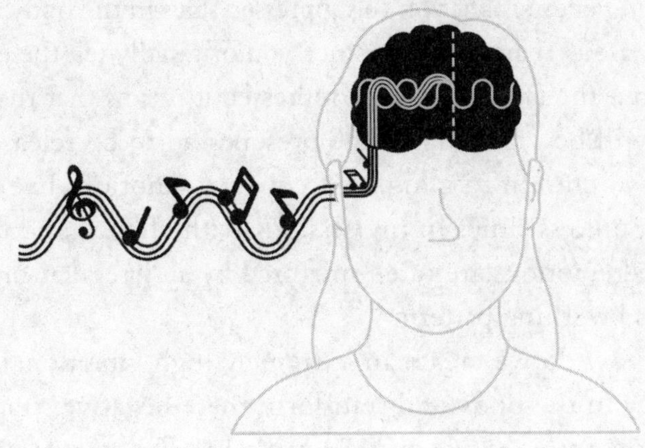

Have you ever wondered how you can tap your foot in time to your favorite tune? Entrainment makes that happen. This phenomenon was first described in 1665 by a Dutch mathematician, physicist and inventor named Christiaan Huygens. He found that when he placed two pendulum clocks on a wall near each other and swung the pendulums at different rates, they'd eventually end up swinging at the exact same rate. They fell into rhythm with one another. He realized that this peculiarity did not just occur with pendulum clocks but appears in chemistry, pharmacology, biology, medicine, psychology, sociology and other fields.

To see how this applies within the body, let's return to our musical example. Music is delivered to our ears via sound waves that travel through the air and into your body. When these waves enter your ears, your brain begins to adapt to the music; it brings its own waves into phase with the music, matching the shape and size of the sound's waves. This allows us to tap along in time.

The idea behind trapped emotions is that this emotional

vibrating energy that you've suppressed stays in the body until it's released. Years may pass since you originally felt the emotion, but the fact that you've repressed it means that the energy will be vibrating in the present and to be released it needs to entrain to a higher-frequency emotion. I've seen this countless times in my sessions with clients, and these trapped emotions are often mirrored by an irregular or disrupted breathing pattern.

So how do we create an extremely high vibrational frequency in the body and transform these negative, trapped emotions into positive ones? How do we release these points of physical or emotional tension?

That's right. Through breathing.

Name of Level	Energetic "Frequency" (Hz)	Associated Emotional State
Enlightenment	700–1000	Ineffable
Peace	600	Bliss
Joy	540	Serenity
Love	500	Reverence
Reason	400	Understanding
Acceptance	350	Forgiveness
Willingness	310	Optimism
Neutrality	250	Trust
Courage	200	Affirmation
Pride	175	Scorn
Anger	150	Hate
Desire	125	Craving
Fear	100	Anxiety
Grief	75	Regret
Apathy	50	Despair
Guilt	30	Blame
Shame	20	Humiliation

All matter, thoughts and feelings have their own vibrational frequency.

INFINITY BREATHING

By now I hope you've become more aware of your breathing, and that you're still regularly practicing breathing slowly with only your nose and diaphragm, in for five, out for five. I hope you've also been using mouth tape, building up to 20 minutes daily, and potentially taping at night to improve your breathing when you're asleep. I want you to continue these functional daily practices. You should also continue to return to techniques like "If in Doubt, Breathe It Out" whenever you're in a difficult situation and don't know what to do, as well as following our techniques for sleep, digestion and pain.

However, if you recognize that you're under constant mental and physical stress—that you're stuck in a certain state—you may need to begin working to break free of emotional addiction and release trapped emotions. In order to do this, we're going to begin a daily dynamic breathing practice called "Infinity Breathing."

This is something I'd like you to dedicate 10 minutes a day to, preferably in the morning, for the next 40 days. This is how long it can take to create the shift.

The name gives it away—it's a free-flowing conscious energy breathwork technique that connects the in-breath and out-breath in a loop, removing the brief pause that comes with healthy breathing at rest.

This way of breathing is not how you should breathe normally, but is an intervention to help shift the energy in your body and alter your biochemistry. The practice can help create a high vibrational frequency to boost entrainment, enabling trapped emotion to complete its natural integration cycle and be processed.

We'll build on this practice in the next chapter. It's important to recognize this is a powerful exercise and its benefits compound over time, which is why I'll be asking you to do it every day over the next 40 days. But for now let's start practicing, so you can begin to get comfortable with it.

EXERCISE 23

INFINITY BREATHING

- Find a comfortable spot, seated or lying down. (Try to avoid the bedroom, because of its association with sleep.)
- Set a timer for 10 minutes, or have some music ready that will last about that long.
- Refer back to your intentions at the start of the book. What do you want to get out of this practice?
- Let your body, mind and breath settle.
- Breathe in through your nose for a count of three seconds, feeling your belly rise.
- Without a pause, release and let your breath go for three seconds, also through your nose. Make sure you're not breathing out vigorously. Just let your breath leave your body without forcing or controlling it.
- Without a pause, breathe in again.
- Keep breathing in this way—three seconds in, three seconds out, continuously, with no pause. Feel the breath flowing in and out of your nose and your belly rise and fall.
- Keep doing this for 10 minutes, until your timer rings.

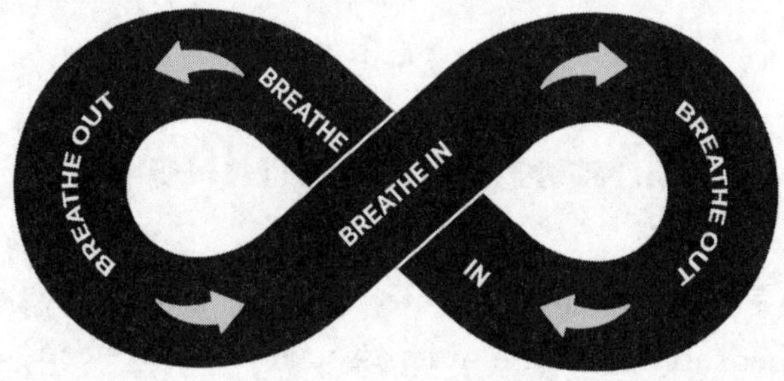

Remember that the Infinity Breathing practice merges the in- and out-breath to form one continuous breath, and that there's no pause between either of your breaths. If it helps, you can imagine your breathing pattern as the infinity symbol, which looks like a number 8 lying on its side. Each half of the symbol is one breath in or out. I sometimes find it helpful to trace the symbol with my finger while I breathe.

Don't worry about getting this right the first time, even if you're the breath-controller archetype. You'll find the flow begins to come with practice. You may start to feel physical changes in your body—energy, tingling, buzzing. You might feel a change in your body temperature. That's OK, don't worry. As you practice this exercise, be aware of any tightness in your breathing cycle. If there's anywhere that feels constricted or where you feel your breathing is not free-flowing, try breathing into that space, or place your hand there and massage gently as you continue breathing. If you start to notice uncomfortable cramping building up, you're breathing too forcefully. So listen to your body and take a break. This practice is dynamic, but relaxed, open and flowing. It shouldn't be uncomfortable.

If you sense an emotion bubbling up to the surface, don't

force it back down. Let it flow. Remember: if you fully feel your emotion and accept it, it will only take 90 seconds to integrate. Emotion may not always come during the practice itself; it might surface outside of it, in your daily life. And that's OK. That's normal.

Infinity Breathing is so powerful because it works to remove trapped emotions, whether or not we know that they're there. It also works regardless of whether we even know that we're troubled by something. It can also help us to overcome experiences we have that influence our beliefs about the world, and heal those emotional wounds that are unavoidable in life, the ones that are written as trauma in our breathing and our bodies.

This will be our focus in the next chapter.

6
RELEASE TRAUMA AND REWIRE YOUR MIND

UNDERSTANDING TRAUMA

I was in Brazil with Scott, my best mate and DJ partner, and Becky, his girlfriend at the time. The sun had just set as we were strolling down a colorful street in Salvador, chatting away. I was holding my laptop under my arm when, from the corner of my eye, I noticed a car rolling up beside us. I turned toward it as the window wound down. My first thought was that whoever was in the car was asking for directions.

Before I knew what was happening, the guy was out of the car, I had a gun to my head and I was being screamed at in Portuguese. Scott was pretty taken aback, but was just about keeping his cool. Becky remained rooted to the spot, terrified by what was taking place. Although I didn't know what the guy was saying, I could guess. I gave him my laptop, wallet and everything I had in my pockets. He turned the gun on Scott, and shouted some more. Scott luckily had nothing on him, but Becky did. She had her bag.

When the gun was turned on her, her nervous system kicked into overdrive and she played dead at the worst possible moment. She was completely frozen still, clutching her bag tightly to her chest. While we were urging Becky to hand the bag over, the gunman was getting more and more frenetic, screaming at us and waving the gun in all directions. For a moment I didn't know what was going to happen. Finally, we were able to prize Becky's bag away from her grip and hand it over. He pulled the gun back, jumped in the car and sped off. Becky burst into tears.

We often get scared by the word *trauma*. We tend to associate it with extreme events and their consequences—soldiers returning from war, violent attacks, car accidents, sexual assaults, near-death experiences, bullying at school or work, or things like our incident in Brazil. The term comes from a Greek word meaning "wound," and like any wound, it can be deep or superficial. It can be caused by one single event or a buildup of multiple events that causes trauma to remain unresolved. Losing Tiff to cancer was traumatic for me, and it took regular breathwork practice for me to fix its aftermath. Being held at gunpoint in Brazil was also traumatic, but I processed it fairly quickly. Although it might sound rather strange, I found it more traumatic to have been falsely accused by my dad of not screwing the lid on the ketchup, when I shook the bottle and it sprayed all over the kitchen when I was seven years old. That took me longer to get over than the robbery in Brazil.

Now, to you, this ketchup incident might not seem that "traumatic." But let's explore the context. I felt that I was being blamed for something I didn't *believe* was my fault, and I wasn't being given the opportunity to tell anybody that it wasn't me. Unconsciously, this appeared to confirm negative feelings that had arisen during earlier experiences. I felt I was being ignored, or that what I had to say didn't matter. What this shows is that trauma can be inflicted not only by something that happens to you, but also when something *doesn't* happen that you believe should. I wasn't asked for my side of the story, and that made me feel like I wasn't listened to. This incident also shows that trauma is highly subjective.

What's traumatic for one person might not be so for another. What's traumatic for you might even be validating or comforting for them. For example, while one person might be scared of ever getting back on their bike after having had an accident, another might think that injuries make you tough, especially if they've been brought up to believe that getting hurt makes them strong or resilient.

Doctor and author Gabor Maté goes one further, and states that "Trauma is not what happens to you, it's what happens inside you as a result of what happened to you. Trauma is that scarring that makes you less flexible, more rigid, less feeling and more defended."[44] This is what I see time and time again reflected in the breathing of my clients.

LINDA

Linda was an extreme example of someone with long-term unresolved trauma. She was in her 50s when she came to see me. She had ongoing mental discomfort—in fact, she said she'd felt unsettled, anxious and prone to negative thinking for as long as she could remember. I could see her breathing was constricted, frozen; there was very little breath flowing in and out, and the little that there was seemed very controlled. I could tell pretty quickly that there was a lot of tension around her diaphragm, especially around the midsection, her solar plexus region. This tends to indicate that the person has a lack of trust or is holding on to fear. Linda was a frozen breather/controller breather.

We began Infinity Breathing to deliberately increase an energetic flow and alter the chemistry in her body. Using hands-on acupressure, I was able to help release some of the muscle tension and contraction in her diaphragm. I also encouraged her to use sound to help vocalize and express any trapped emotion. This is something I find useful to help clients dislodge particularly stubborn trapped emotions. We've all felt the release when screaming into a pillow. We'll discuss this in more detail later.

After about 20 minutes of breathing, I could see a lot of energy build up, followed by an outrush of emotion. Tears began rolling down her cheeks, the tension in her solar plexus seemed to vanish, and her breathing became open and free-flowing. After our breathing session, she shared with me how she'd been deeply traumatized by a car crash that had taken place about 40 years earlier. She'd been tense ever since, unable to let the experience go and incapable of getting into a car. It had affected her relationships, her career—every aspect of her life. After the session she said she felt liberated and alive again.

When a traumatic event such as Linda's car crash occurs, or something happens that's perceived by an individual to be traumatic, the memories become stored in the brain and nervous system in a maladaptive way, keeping you trapped in the trauma. Your breathing becomes rigid, moving you into one of our archetypes, and your emotions are frozen rather than processed. Our Infinity Breathing practice opens your breathing and creates an energetic flow in your body. It helps release tension and any negative emotional charge present, enabling you to store your memories in an adaptive way. A person can then look back on an event without being triggered and feeling a lot of distress.

My memories of Tiff and her death are still with me. But the crippling feeling of grief is no longer there. When I was experiencing grief, my unconscious mind stopped the flow of breath into my chest in order to protect me from feeling emotional pain in my heart. But by releasing the tension blocking me from deeply feeling, and opening up my body to the full flow of my breath, I could finally process this trauma and begin to move forward. Yes, every Valentine's Day, the anniversary of her death, my mind and thoughts can produce more emotion. But because my trauma has been processed, my breathing is open to flow, and that emotion no longer impacts upon my body and mind in the same way. It comes and then it goes. I like to think of it as like an old ankle injury. Sometimes when you're dancing, you feel a slight twinge. But it doesn't stop you from being on the dance floor. When your breathing is open and flowing, trauma doesn't have to define your happiness.

Trauma doesn't have to define your happiness.

OPEN YOUR HEART

An open heart is a state of being where you feel free, accepting and expansive. Your breath is open and flowing without obstruction and you experience a sense of love, joy and trust. We all long to experience an open heart, but at times we unconsciously feel too scared and vulnerable to reveal ourselves in this way. Grief and other forms of trauma can cause our unconscious mind to close down our "heart center" in our chest and prevent our breath from expanding into it. This is its way of deflecting our hurt and pain by shutting down and pretending it doesn't exist or matter. This makes it more difficult to connect with ourselves and others in a loving way.

SIGNS OF A CLOSED HEART:

- Feeling emotionally paralyzed, numb or stuck with life.
- Being angry, mean and cynical, or critical of yourself and others.
- Withholding love and affection toward yourself and others.
- Finding faults in people and judging others.
- Avoiding new experiences and adventures.
- Avoiding connection with others.

SIGNS THAT YOUR HEART IS OPEN:

- Smiling, laughing and being expressive.

- Demonstrating affection and kindness toward others.
- Focusing on the good in people rather than the negative.
- Being open to new experiences and opportunities.
- Being resilient through adversity.
- Feeling positive, peaceful and energetic.

EXERCISE 24

BREATHE TO OPEN YOUR HEART

Only try this when you are confident that you are breathing with your diaphragm and that your breath is expanding into your lower torso before anywhere else. See if you can then start to have this secondary expansion through your midsection and up into your chest, flowing like the movement of a wave. Belly-midchest, release and repeat.

- Find a comfortable position, seated or lying down.
- Place one hand on your heart and one hand on your belly.
- Breathe in through your nose, expanding into the hand on your belly.
- On this same in-breath, feel an expansion in your midsection between your hands, like a wave flowing upward, until the final expansion into the hand on your heart.
- Breathe out, through your nose—relax and let go.
- Repeat 10 rounds.

You may wish to practice this, repeating the mantra: "I choose to open my heart," or "it's safe to open my heart." You can also focus on this fuller flow and expansion of your breath in your Infinity Breathing practice, especially if you feel your heart is closed.

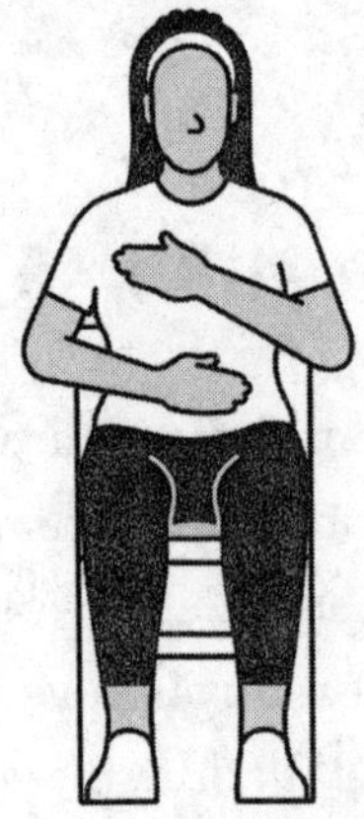

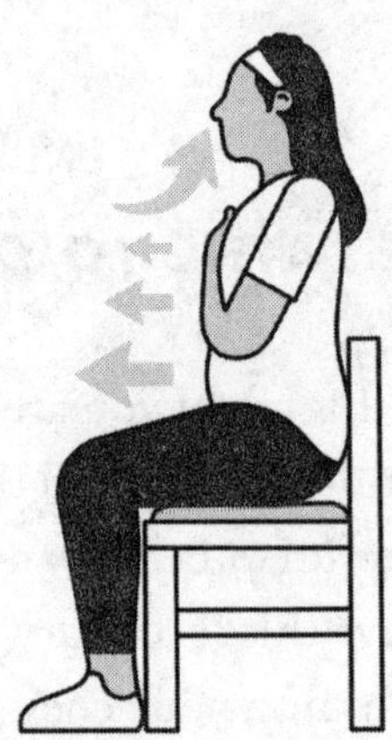

BELIEFS AND BREATHING

As the incident with the ketchup shows, trauma can get tangled up with your beliefs, and your beliefs can have their origins in the experiences you have, whether good or bad. Maybe the time your creative project was praised by your teacher has created a lasting belief for you, or maybe hearing that you were a great child a lot at home has cemented the belief that you're a good person. Maybe at times your needs were not met in a way you believe they should have been. Maybe, like Jane and Anna from Chapter 1, you've also formed beliefs that certain things are dangerous or safe based on what you were exposed to when you were a child. You might believe you're not good enough because of something you heard someone say about you when you were young, or because you were always struggling to meet someone else's standards. The beliefs that these experiences form are validated and reinforced over the years by repeated thoughts.

Let's return to Jane and Anna for a moment, as it's important to understand that there are different types of experience that affect your breathing and behavior. This basic example can be substituted for any experience that evokes strong emotion. If, like Jane, you were bitten at a young age by a dog, the shock of the experience would force your breath pattern to change, new neural pathways in your brain would form, and you could develop the belief that dogs were not safe and that you should avoid them in the future. This is *felt* experience.

If the dog had bitten my mum, however, rather than Jane herself, and my mum then told Jane never to go near one, Jane would still learn that dogs were not safe to be around and would therefore avoid them. Even though she hadn't personally experienced the dangers of dogs, this would constitute a *learned* experience.

All our experiences shape what we believe about the world. Felt and learned beliefs can spread far and wide, becoming *shared* beliefs. Let's take another simple example—everyone living in an apartment block might be afraid to use the lift if their neighbor had been stuck in it for three hours and told those around them that it wasn't safe. We commonly see *shared* beliefs circulating in families, communities and wider society. Imagine if you turned on the news one day and were told that not a single lift in the world was safe to go in. A single belief can spread like wildfire.

There's also another way that we can form beliefs—through repeated thoughts—and these can be passed from generation to generation. The culture in which we grow up, the language we speak, the information disseminated through various forms of media and our surrounding environment can all color our belief system.

For example, women are sometimes taught in certain cultures that it's OK, or even expected, to be more emotionally expressive than men. In some circles, men showing emotion (except anger) is still thought of as "taboo." This was true for me, my dad, his dad, and his dad, possibly for generations before him, back until the moment when it first became culturally unacceptable for men to *really* talk about how they felt. I experienced this as a little boy when I opened the front door and found my dad crying.

"Go get Mum," he said. "I've fallen off my bike."

Men in the kind of society I grew up in never cried unless they were badly hurt, and even then it was often frowned upon. The truth was that he was having a nervous breakdown due to work stress and burnout, but he felt he had to project an image of strength to his son. And in doing so, he projected this belief onto me, or at least added weight to a shared belief that was common in my community.

So I grew up believing that men should be "strong." That could have come from my dad, but it could also have sprung up from a culture where "big boys don't cry" was something I heard a lot. This stereotype was certainly reinforced by Tough Ted and the years I spent doing judo, which I started at four years old. My first-ever judo coach even had us call him "Superman."

EXERCISE 25

INFINITY BREATHING WITH SOUND

Sound and movement can help accelerate the results of your Infinity Breathing practice. All sound, regardless of whether it has a high or low frequency, has a higher vibration point than solid mass; this means that making sound causes your body to begin to vibrate, and promotes those low resonating, negative emotions to start to clear through the concept of entrainment. Movement can release physical tension, and so further contributes to this sense of clearing. So I'd like you to now add some sound and movement during your daily Infinity Breathing practice.

- Find a comfortable spot, seated or lying down. (Try to avoid the bedroom, because of its association with sleep.)
- Set a timer for five minutes, or have some music ready that will last about that long.
- Refer back to your intentions at the start of the book. What do you want to get out of this practice?
- Take a moment to settle, and anchor yourself and your breath in your body.
- Breathe in through your nose for three seconds, feeling your belly rise.
- Without a pause, release and let your breath go for three seconds, also through your nose. Make sure you're not breath-

ing out vigorously—just let your breath leave your body without forcing or controlling it.

- Without a pause, breathe in again.
- When the timer rings, drum your hands on your knees while making a humming sound.
- Repeat the drumming and humming for three rounds.
- Return back to the Infinity Breathing practice for another five-minute round.
- You may feel the tingling and buzzing feeling intensify through this process. Remember, you're in control, so stop or pause if you feel uncomfortable. If emotion bubbles, let it flow and let it go.

THE BRAIN OF A CHILD

Early childhood is not just a period of physical growth; it's also a time of mental development, involving changes in anatomy, physiology, chemistry and the nervous system. Between the ages of zero and seven are our programming years. During these years we're almost in a permanent state of hypnosis, being conditioned by the environment around us, open to suggestion. Our brain is in a super-absorption learning state, memorizing, reasoning and problem-solving, and we're forging our relationships with ourselves and the world around us.

Our brain development is rapid at this time. In some ways, our childhood years are when we download our files, operating system and programming. These are the years when we begin to develop our core beliefs—the beliefs that influence how we see ourselves and our world, the beliefs that play a decisive role in our physical and mental health in adult life. And although these beliefs aren't necessarily *objectively* true, they provide a framework for interpreting information. You could think almost anything and practice, practice and practice that thought—but the belief that the thought will transform into is just a story you've told yourself.

BRAIN WAVES

At different periods in our lives, different types of brain wave dominate, and we move between them as children and adults. This can happen by itself, as when we're in deep sleep, or we can consciously use tools like breathwork or meditation to enter different states and promote a change.

There are four different types of brain wave, which are differentiated on the basis of their electrical patterns and frequency:

- **DELTA:** Children: 0–2 years old (and even in the womb). Adults in deep sleep. This is a period of huge development in children, and these brain waves are associated with the deepest levels of relaxation and restorative healing in sleep.
- **THETA:** Children: 2–5 years old. Adults in hypnosis or just before nodding off. It's the realm of imagination and daydreaming. A super-learning state in which we're open to suggestion.
- **ALPHA:** Children: 5–7 years old. Adults in meditation. We're peaceful and relaxed, though alert. We can draw conclusions from the environment, and the inner world of imagination tends to be as real as the outer world of reality. This is a great state to be in when brainstorming or learning.
- **BETA:** Children: 8–12 years old and onward. Adults spend most of their time in this frequency. It's the world of conscious, analytical thinking. The mind is awake, focused, alert and capable of logical processing.

Our childhood years are when we download our files, operating system and programming.

CHILDHOOD PROGRAMMING AND CORE BELIEFS

Let's delve a little deeper into beliefs—specifically, your core beliefs. Your core beliefs are the ideas about the world that your brain forms when you're a child, and they remain unconscious throughout your life. They make up your childhood programming, which is based on your experiences, the things you see people do, and the advice others give you when you're growing up. In adulthood, they will show up in your thoughts, feelings, perceptions, and will play a part in your actions, the decisions you make, how you choose your friends and partners, and so on. They're products of your conditioning by your experiences and surroundings, not objective facts.

It's through this conditioning that you make sense of the world. This interpretation of the world in the light of our existing beliefs is what we call *confirmation bias*. A part of your brain, the reticular activating system (RAS), filters out a lot of the information it receives so that we're not overloaded with stimuli. In this filtering process, it prioritizes the things that it thinks you need to recognize, like something you have been focusing on or thinking about, or something that is part of your deeper programming, like your core beliefs. For example, if you believe dogs are unsafe, your brain may filter out their friendly behavior or interpret it according to that belief. You might perceive their playful jump as a vicious attack. Or, if you believe "no one listens to me," then you may filter a conversation so that you hear or interpret something in a way that backs that up. What the thinker thinks, the prover proves. We hear what our beliefs want us to hear, we see what our beliefs want us to see—this confirmation bias means you'll see evidence all around you of something you already believe, "proving" that your belief is grounded in reality. This even happens through focused intention. Have you ever noticed that sometimes you'll think about something you never usually think about, only to then see it everywhere you look? It's why the positive intention setting we did at the start of the book can be so powerful. We are starting to reprogram your RAS so it works for you, not against you.

In some ways, your core beliefs give your RAS its operation manual. Which then works as your manual for interpreting what happens in your life. We'll have had part of this manual for so long that it feels like part of our identity, and that makes it challenging to alter. But if our manual is the reason for our continual self-doubt, our insecurity, our low moods and constant desire for approval, then we owe it to ourselves to challenge it and change it. Maybe you're programmed with beliefs like "Life's not fair." "Why does it always happen to me?" "I'm not enough." "If I just have this or that then I'll be happy." "I'm not deserving." "I'm worthless." "I'm unlovable." Behind these surface-level characteristics that we may think of as being part of our identity lies a more authentic, creative version of ourselves—a version we need to reconnect with. This is all mapped in your breathing and the breathing archetypes you experience.

To change your behavior, you must replace these old unsupportive beliefs with new, helpful ones. As always, this change starts with awareness. If you don't question or examine who you think you are, alongside your breathing, then you may get stuck with old beliefs that could limit you from feeling truly happy, alive and free. You deserve to be happy, even if you do not think you do.

We all come from different backgrounds. We are conditioned by surroundings, our peers, teachers, role models, and, most of all, our parents and caregivers. I like to think most parents and caregivers will be doing their absolute best to provide you with everything you need in a way they know how—but this isn't always the case. Parents and caregivers act according to their own core beliefs, learned from their parents, their schooling and significant other people, and of course from their surroundings, experiences and own unresolved trauma. This can influence the ways in which they act. There will be times when they're under great stress, shut down, when in the heat of the moment they'll say things or act in a way that's hurtful. If you are experiencing a lot of distress because of something that happened to you when you were a child, please seek additional support alongside doing the practices in this book. Speak to your GP, a therapist or a trained professional who will have some brilliant resources and ways to help.

Maybe your childhood was lacking in something. Maybe you couldn't express yourself, so your core belief is that you're unheard or unseen. Maybe you felt you had no one with whom to share your concerns about school, your friends or home life. So your core belief is that your emotional pain is not important or that your feelings are not trustworthy. Maybe the adults in your early life were pushy, trying to live their life through you, so your core belief is that it's absolutely necessary to succeed and that your authentic self doesn't matter. Perhaps you heard throwaway comments like "Your sister is smarter than you," "You're so clumsy" or "You'll never make it onto the team." Whatever you hear or experience growing up can weave itself deep into the fabric of your mind, create tension in your breathing cycle and have a huge impact on your adult life. It creates the manual that your RAS operates from today.

Even if your childhood was very pleasant, you still have core beliefs. You still download your manual to help you make sense of the world. But if you keep running into the same problems over and over again, it's probably time for you to update your manual.

Let's take a look at mine. We have already explored the entry around the notion of strength—reinforced through experiences, ideas of traditional masculinity, judo training, Tough Ted and so forth. But let's unpack some others. Growing up, I was fortunate to have supportive teachers, good friends and a stable home environment. I'm very grateful for my parents, who worked hard and dedicated all their time and energy to create a positive and healthy upbringing for me and my three siblings. However, I still ended up with some unhelpful entries in my manual, although the process was subtle. For example, there was an entry around money: *Money doesn't grow on trees*, my parents would say, probably because they worked so hard as a tag team to make enough. My dad would work crazy long hours as a catering manager to provide for us; my mum too: she spent her nights working as a nurse, a job she juggled alongside managing their bed-and-breakfast side hustle. She'd come in from her night shift when we woke up, make guests their breakfast, take us to school, get some sleep, then pick us up from school, taxi us to our various hobbies, put us to bed and head back to work. They wanted the best for us, and to provide us with everything they wished they had growing up. They had their manuals too. They had their bag of bricks, their core beliefs, their conditioning. Everyone does. They went through some very difficult experiences. So, growing up with hardworking, often stressed parents who had seen more than their fair share of tragedy, it's no wonder that in my mental manual, there was the notion that you've got to work hard to achieve in life. In many ways, this is a great belief to have, but it did reinforce the idea that you must always be "doing," always on, always working. This is probably how I got addicted to stress and had a stressful breathing pattern. It might have even been the birth of Airport Stu.

As a family, we were close. Yet, growing up, the words *I love you* were rarely uttered in our house. Perhaps they were even frowned upon. My parents felt "I love you" lacked meaning when used casually. They believed that the best way of saying "I love you" was by showing it through what you do. So, in their mind, they didn't need to use this phrase to prove their love for us. And although I never felt unloved in any way, not hearing those words developed into a belief about how *I* should express love in my relationships as an adult. As I got older and began dating, I noticed feeling uncomfortable if a girlfriend said "I love you," as my childhood programming told me that this was not necessary. You could argue that if I had heard *"I love you"* every day, but from parents who didn't actually show it, it would have had a very different impact on my relationships as an adult. This shows that, no matter what, we are all conditioned. We all have our manual. And it affects how we operate today.

Once you become aware that confirmation bias is something that we *all* suffer from, you can try practicing being a little more open to your perception and judgments. When you know that your breathing influences your thoughts and feelings, you can begin to use it as a tool, alongside some conscious awareness and reprogramming, to change your beliefs. And that means you don't have to spend your life shackled to beliefs formed by your experiences in youth. You can free yourself. So it's time that we start unpacking our bag and changing our manual—but first, let me introduce you to Jasmine.

**What the thinker thinks,
the prover proves.**

JASMINE

Jasmine joined a group session I was leading in London. As part of the practice, I asked everyone in the group to make "toning" sounds and move their body at various times during their breathing practice, similar to our "Infinity Breathing with Sound" exercise, as a means of moving any trapped emotion.

A short way into the session, I noticed a woman who was barely making any noise at all. I went over and encouraged her to go for it. But still, nothing came out. We continued in this format for 5–10 minutes of breathing, followed by two minutes of movement and sound. Over the course of the hour-long session, she slowly opened up, making more and more noise each round, until eventually she was the loudest person in the room. She was more than loud, and ended up crying with laughter and full of joy.

At the end of the session, she came over to thank me. She said she'd dreamed of being a singer as a child but that someone told her that her voice was terrible, an extremely traumatic experience for her. Believing herself to be terrible at the thing she loved to do most had badly dented her confidence and made her withdrawn, unhappy and quiet. The breathwork class opened her up, enabling her to overcome that painful experience as a child. She began to believe she had a right to express herself. She messaged me a month later to say she'd started singing lessons and was loving them.

EXERCISE 26

ARE YOU A PEOPLE-PLEASER?

Children who grow up in chaotic or controlling environments can become people-pleasers, learning that by being good or behaving well, they can avoid conflict and attract love. This is called the "fawning" response, an alternative to both fight or flight and freeze. As adults, people-pleasers can be emotional shape-shifters, minimizing their own needs and wants to satisfy those of others. This behavior stems from a belief that love is conditional.

Are you a people-pleaser? Take my quick quiz:

- Do you struggle to feel like others "see" you?
- Do you find it hard to say no to people?
- Do you unload your emotions onto strangers?
- Do you feel guilty when you're angry at other people?
- Do you feel responsible for other people's reactions?
- Do you compromise on your values for the sake of other people?
- Do you disconnect emotionally from social situations?

If you answered yes to any of the above, you might be a people-pleaser. People-pleasing isn't always the result of trauma, and most of us have some people-pleasing tendencies. The important thing is to recognize that we can never really know what other people want, so our attempts to please can sometimes backfire. If we truly want to be liked, we simply need to be our authentic selves.

It's time that we start unpacking our bag and changing our manual.

EXERCISE 27

WHAT'S IN YOUR BAG?

Since core beliefs often relate to things we experienced as a child, especially during those programming years from the ages of zero to seven, a useful exercise is to think about what people frequently said or did while you were growing up.

Think about your childhood, particularly from the time of your earliest memories to seven years old. Here are a few questions to help:

- Were there any traumatic experiences, big or small, that still impact your behavior today?
- Did your caregivers pay you a lot of attention?
- How did they show you love?
- What messages did you receive a lot as a child? Were you told to shhh and not to cry? Were you told off for something that you didn't do or told that you weren't good enough at something? Were there messages around money, work or relationships? Were you taught that people could be trusted?
- How did your parents treat each other?
- Were they at home most of the time or out?
- Was there something you were told you weren't good at?
- Was there something you were told you *were* good at?
- Did your caregivers make judgments about others?

- Did anyone say that you weren't allowed to express an opinion?
- Did your caregivers complain a lot about work?
- Do you have siblings? What was your dynamic with them? Were you constantly being compared with them?

These questions won't cover all the things you might have heard as a child, but they'll certainly get you thinking. Now take a moment to consider those messages you received. Can you think of any ways in which they might be influencing your behavior today? Can you start to change your core beliefs to ones that are more positive, or encourage your personal growth? Check the table for some inspiration.

Please note: If you can't remember anything at all from your childhood, it may be the case that you were in a constant state of "freeze" during this time. This shut-down, withdrawn state is a form of psychological protection from experiencing something traumatic or overwhelming. This can still be processed through our "Infinity Breathing" practice.

CHANGE YOUR CORE BELIEFS

NEGATIVE BELIEF	REFRAMED THOUGHT
I'm a failure/ I'm not enough/ Everything is my fault / I never do anything right.	I'm doing my best. I am worthy. It's safe to let go. I learn from my mistakes.
I'm unlovable / undesirable / unattractive / ugly.	I love and accept myself unconditionally. I am enough.

NEGATIVE BELIEF	REFRAMED THOUGHT
What's the point? / Why do I bother?	I have a positive contribution to make to the world. I follow my intuition.
I'm just an anxious person.	I am safe. I choose to be confident. Anxiety does not define me.
I'm always broke. I never have any money.	There are opportunities everywhere. I have the skills to earn a living.
I'm stupid.	I trust and believe in myself. I learn from my mistakes and get better.
No one listens to me.	I am heard, I am seen. I express myself openly and freely.
Why does it always happen to me?	What is this teaching me?

Now, there's one other influential experience we all have that we need to discuss. It lands first in everyone's bag and might become the first entry in your manual.

BIRTH TRAUMA

When you were born into this world, it came as a bit of a shock, however you arrived here. Birth is one of our first felt experiences. It goes in the bag—there's no avoiding it. It's traumatic. And not only is birth traumatic for you, but it will most likely have been traumatic for your mum, and probably your other parent too. That first breath may play a big part in shaping who you are and how you perceive reality. It's the first entry in your timeline.

Some have even suggested that the way in which you were born—induced, breech, by cesarean section or another way—can correspond to beliefs you form about the world. Babies who were induced might, according to this hypothesis, grow up to believe that they can't do anything on their own. If it was a difficult birth, then you might grow up to think the world is not a safe place to be. For some, that fear may be further reinforced by early childhood experiences, such as being in a chaotic home environment. It's a belief that may play out throughout your life in the feelings you have, the choices you make, and it may lead you to become a stubborn archetype like a reverse breather or even a breath grabber, distracting yourself from being here, now. Your unconscious mind may put you in a constant state of fight and flight or freeze. Maybe you feel permanently anxious, stressed or socially awkward.

I was first introduced to the concept of birth trauma when I came across the work of Leonard Orr and his practice of "rebirthing."[45] Orr is considered one of the pioneers of the breathwork movement in Western culture. His teachings are said to have been inspired by Kriya Yoga's guru Mahavatar Babaji (Yogananda's guru—there's that book again!). Yet Orr's first insight, and his ability to unlock and resolve his own birth memories, came to him in 1962, though not from the Himalayan ashrams or spiritual scriptures.

His first major breakthrough came to him like many of our *Aha!* moments do: in the bath. He found that by merging the inhalation with the exhalation in a gentle relaxed rhythm—just as we do when we do our Infinity Breathing—people could, as he described it, "heal the eight biggies of human trauma: birth trauma, the parental disapproval syndrome, specific negatives, the unconscious death urge, karma from past lives, the religion trauma, the school trauma, and senility." That's a pretty bold statement to make from a bathtub.

Orr and his teachings spread like wildfire across the Western world. He's believed to have taught over 100,000 instructors and is estimated to have reached over 10 million people. There have been good things said about this technique and some not-so-good things, with some rebirthing extremists taking things too far. However, rebirthing is still practiced today and its technique is the basis for most of the common "conscious energy" breathwork schools recognized today.

When you're training in anything that edges into slightly spiritual territory, you're often told stories of the man who walked on water or the guy who did so much yoga he started levitating or the person who healed all their troubles and ailments by breathing air. Although my skeptical alarm bells rang time and again as I explored the world of breathwork, the exercises and practices I was learning were having a profound effect on me and many others I worked with, met with, trained with and shared stories with. However, the notion of birth trauma—the practice of rebirthing techniques—was something I only ever learned in training. And then I met Danny.

DANNY

I was introduced to Danny by his mum, Trish. She told me he was struggling with severe separation anxiety. It was so serious that he'd panic if he was playing in the garden and she shut the back door. She wondered if it had something to do with the difficulties she'd had when she gave birth to him. I told her that I'd only ever learned about birth trauma, but that I was very happy to see Danny and that I hoped I could help in some way.

The first time I met him, his body language told me he was too terrified to engage in conversation. His shoulders were hunched and his body was turned in, collapsing his breathing. He stood behind his mum's big hair and looked down at his Nikes. Rooted to the spot, his posture was frozen and his breathing paralyzed.

Mentally, he seemed shut down and spaced out—the hallmarks of a breather who is both collapsed and breathing in reverse. In fact, his breathing was almost nonexistent, and every so often he'd take a large gasp as if to catch his breath. People who live in a constant state of fear, as Danny did, find it hard to think straight, and any demands to perform can lead them to shut down further. So I took things slow. I sat him down and prompted him to breathe. I soon could confirm he was a reverse breather, so I encouraged him to take deeper breaths into his torso as I breathed along with him.

Over four sessions, Danny's breath deepened, his diaphragm

started to engage, his posture improved and an even color returned to his cheeks. Week by week he seemed a little more confident and more willing to talk. In the fifth session we had a breakthrough. After about ten minutes of Infinity Breathing, he erupted with emotion, yelling at the top of his lungs for over a minute. And like magic, his breathing reset. It was expansive, open and flowing, efficient and effortless.

The next time I saw Danny he came into my studio without his mum. He was engaged and charming, and told me he'd joined the school football team. He seemed like a completely different person. He appeared freed from his trauma, although whether it took place at his birth or later on in his life, we'll never know. But it was one of the most dramatic transformations I'd ever witnessed.

I'd like you to revisit our "What's in your bag?" timeline exercise (Exercise 27). Could the person you were growing up and the person you are today have been impacted by your birth experience? You may need to ask a few people, of course, about what your birth was like, but it could be another clue to help you let go of deep-rooted beliefs that no longer serve you.

For most of us, although our birth experience is traumatic, the love, care, food, warmth, protection and connection we receive from our parents is usually enough to allow us to overcome the trauma of our crash-landing arrival on earth. We come to the conclusion (or belief) that it was not so bad after all. That cleans out the bag, so it's nice and empty when we continue on our life journey. It's like as an infant you're a blank space of awareness, and your brain is a sponge, absorbing information, ready and wanting to make regular connections and pathways to help you navigate life.

But not everyone is lucky enough to be born into and grow up in loving, caring homes. Sometimes caregivers are still dealing with their own trauma and are too shut down to attend to our needs. These experiences and influences can affect your life today, but you can take back control.

You can't change your past, but you *can* learn from it so that you can make better decisions now. This is about understanding and making peace with people and your experiences, learning from them and being curious about yourself, your beliefs and your behavior, not eradicating them from your history.

You now know that through daily awareness and a regular functional breathing practice you can begin to put down your bag of bricks, alleviating your stress and easing any pain. But you can also begin to take out what's inside that bag of bricks using Infinity Breathing. You can start to resolve any trauma, even if that trauma is so tightly intertwined with your beliefs that you can't recognize it for what it is.

EMPTY YOUR BAG

We've learned in this chapter how we're almost all carrying around trauma, big or small, and how this can be entangled with our core beliefs: the ideas we were exposed to when we were children that we've now internalized. We've seen that these beliefs can even be influenced by our birth and we've also explored how they are like a mental manual, telling us how to find our way through life. But we can overcome harmful beliefs and experiences through breathing. We can begin to empty our bag of bricks.

Sometimes, our core beliefs and behaviors are so deeply entrenched that it's difficult to free ourselves from them immediately. It takes time and dedication to your breathwork practice. When we're really trying to unlearn habits and behaviors that are making our lives difficult and preventing us from moving forward, we have to embrace a daily practice. That will allow us to hit the reset button. Part of that reset is also conscious reprogramming—rewiring the story we tell ourself, changing our manual and core beliefs. And you can do that through affirmations alongside your daily Infinity Breathing practice.

EXERCISE 28

INFINITY BREATHING WITH SOUND AND AFFIRMATIONS

I would like you to add some affirmations to your "Infinity Breathing with Sound" practice. These can be the three statements you chose at the start of the book, or something new, to help rewire your core beliefs. For example: I am enough, I am powerful, I am worthy of receiving everything I need. Anything that feels aligned to you.

So using the same practice format as Exercise 25 (Infinity Breathing with Sound):

- Find a comfortable spot, seated or lying down.
- Set a timer for 10 minutes, or have some music ready that will last about that long.
- Allow yourself time to anchor into your body.
- Notice your body, let it soften. Let go of any tension in your face, your jaw, your neck.
- Notice your mind, be aware of your thoughts, judgments and opinions.
- Now feel your heartbeat in your chest and say your affirmations, your three statements.
- Start your breathing.
- Breathe in through your nose for a count of three seconds,

feeling your belly rise first, then flowing and expanding upward like a wave, through your midsection and chest.

- Without a pause, release and let your breath go for three seconds, also through your nose. Make sure you're not breathing out vigorously—just let your breath leave your body without forcing or controlling it.
- Without a pause, breathe in again.
- Continue with this Infinity Breathing flow.
- Whenever you feel the desire, drum your hands on your knees while making a humming sound. Do this for three rounds before returning back to the Infinity Breathing practice, and repeating until the 10-minute timer is up.
- Settle here, anchor back into your body, fully relax.
- Slow your breath down.
- Feel your heartbeat again, maybe even place your hands on your heart.
- Feel the appreciation for your heart beating, for the life and vitality in your body.
- Feel gratitude for all the things that make you feel safe and loved—the people in your life, the resources and opportunities you have available to you. Sometimes this brings a flow of emotion and that's OK—let it flow.
- Come back to your heartbeat and now repeat your affirmations, whatever they are: *I am strong, I am loving, I am supported. I am proud. I choose to be me.*
- And don't just say them in your mind, really feel them in your body.

- Before coming back into your space.

Remember these affirmations are unique to you and will help you start to change your manual and core beliefs.

7
LET GO AND TRANSFORM

THE PERFECT MENTOR

All throughout my breathwork journey, my beliefs about what breathwork can do have been tested and validated through my experience and the experiences of my clients. I've seen people move on from grief, overcome trauma and become happier, healthier, more functional people almost overnight.

I know from my own experience that breathwork can fix these problems. But I also recognize that if I want to help as many people as possible, I need to be able to explain these transformational experiences in the language of Western science. More research and study into breathwork are needed to prove why these happen, and I'm determined to play my part. A few years after beginning breathwork, I started my search for answers, and I found a mentor—someone with a background in science and research—to help me. And I met him, of all places, in Ibiza.

I've always felt a deep connection with Ibiza—the nature, the sunsets, the culture—and I've spent many magical summers on the island. Geologists say it's one of the most magnetic places on earth and I sometimes wonder whether that's why I found my perfect research partner there.

This trip was a bit different from the ones I was used to. It was my first time in Ibiza as a health practitioner, not a DJ. I'd been invited to deliver workshops at the annual International Music Summit. They wanted to create a space for the electronic music industry to understand mental health and explore support strategies.

I was a little groggy from an early start. The sun was just rising as I clambered into the back seat of the minibus shuttle that would take us from our hotel to the conference venue. The cab was full of experts from around the world but I was new to the game, and if there was ever a time to feel a surge of impostor syndrome, this was it. There were sleep experts, neuroscientists, world-class coaches. I felt a little out of my league, to say the least. I noticed my breathing had frozen. Sitting in front of me on the bus was an older gentleman in a cream Panama hat. He wore round specs and had an aura of intelligence. I hadn't quite woken up when he turned around and began to interrogate me.

"Hello, Stuart. I'm Norm. I've heard people talk about this powerful breathing technique you teach. What's the scientific research behind this technique?"

Gosh. Straight in for the kill with a question I'd been trying my best to answer myself over the last year. I did not know at that point that Norm was Dr. Norman E. Rosenthal, MD, a world-renowned scientist, researcher and psychiatrist who had led the team that first coined the term "seasonal affective disorder" (SAD) and pioneered the light therapy widely used to treat it. A bestselling author, he'd worked with a long file of A-list clients. He knew his stuff.

There was a long silence. It was one of those tumbleweed moments. I wished I had a solid answer. *Give me a break*, I thought. *I've just woken up*. But I answered with:

"It's a bit patchy."

Dr. Norm gave me "the look." His well-trained eyes stayed still, but he might as well have rolled them and tutted at me. He didn't say a word, but what I heard, as if by telepathy, through my imposter lens, was "That's fucking ridiculous, and useless. Why are you even here?" As the minibus started to move off, he turned around to face forward.

Throughout that day, without my knowledge, Dr. Norm observed me from afar as I ran my group breathwork sessions. He also conducted some further inquiries by asking some participants about their experience with me. I had no idea that I was quietly being investigated. Later, after we'd both finished our various workshops, lectures and talks, he approached me and said:

"Your obvious commitment to your practice and your sincerity have elicited my curiosity. I wondered if you might have time to give me a private session?"

I smiled and replied, "I'd love to."

The following morning, just as the sun was waking, we met in a treatment space in the grounds of the hotel. Before we started the session, I explained that we'd be working with powerful and dynamic conscious-energy breathing practices that could release trapped emotion and trauma, and help to rewire the brain. Norm placed his hand on my shoulder and gave me the look again.

"I've been a therapist for 40 years," he said. "I'd be very surprised if anything came up."

And so we began.

NORM

Norm's session was like many other sessions I've run, only this time it felt like the stakes were higher. I could tell he was skeptical, and I wanted to prove the value of what I was doing. As I guided him to breathe, energy started to flow, he twitched and shook, and after a while I could see he was starting to process something. There was tension in his right shoulder, which, if you take a look at our body map from Chapter 5, suggests unexpressed anger.

I helped to release it physically, through applying pressure and light massage, yet his shaking grew more and more intense. There was redness in his cheeks; anger was moving and building, and he was starting to make a noise. The anger seemed to work its way into his jaw in the form of tension, twitches and spasms, and finally there was an explosion of emotion, a yell, before tears began to flow freely down the sides of his face. It went on for almost a minute and a half, precisely the length of time it takes to process emotions fully.

We kept breathing, but now the emotion changed; tears were replaced by laughter—proper belly laughing, the way you'd laugh if you'd just heard the funniest joke ever told. He laughed so hard that my energy entrained to his, and I wanted to laugh with him. Again, this lasted for 90 seconds before his body relaxed. His breathing became open, flowing and effortless. I took him back to a more relaxed state. When it was over, I passed him a glass of water.

"How are you feeling?" I asked.

"That was very…" There was a long pause. "…interesting."

That was all he said.

"Anything you'd like to share? Would you like to talk about your experience?"

"No. It just was just very…" There was another long pause. "…interesting."

He collected his things and went off to his room.

I felt vindicated by Norm's experience, but I didn't hear from him that day as he took an early-morning flight back to Washington, DC. About a week later I got a call from a US number. It was Norm, and he proceeded to tell me what he'd experienced.

Emotions that he thought he'd processed years ago had bubbled up to the surface and come out, and ever since the session, he said, he'd felt unusually positive and at ease. Clearly, as with so many of us, the emotion he thought he'd dealt with was just locked away in the cellar of his unconscious, unseen, unheard, but affecting his daily life. But Norm had something else to say.

"The excitement I'm feeling in response to our breathing session reminds me of how I felt when my colleagues and I first started to explore seasonal affective disorder and light therapy in 1981, or when I came across Transcendental Meditation."

This is something he tours the world lecturing on. "The research on your technique seems thin. I'd love to help you fill in the gaps."

WHAT WE FOUND

Since that conversation, Norm has guided me to gather the testimonies of 636 people, and the results continue to shed light on the impact of breathwork on people's lives. Although my mission was interrupted by the COVID-19 pandemic, we've continued to gather data from clients who attend my online sessions.

So far, the self-reported effects of breathwork have been overwhelming. The same stories come up time and again: improved mood and well-being; release of long-held trauma; improved ability to move forward; and new insight. Across men and women, groups and individuals, even online and offline (which I found surprising), the results were consistently positive. Yet we didn't want to kid ourselves about the results, so we asked participants to report any adverse effects. Some people—19 percent—said they experienced these, including physical discomfort, heavy emotions, feeling tired and withdrawn. However, even though some felt adverse effects, 99 percent of all participants reported that the experience was helpful, with 26 percent of these indicating that it was transformational.

It was a promising start. But there's more we want to learn. What has become clear during my breathwork practice and teachings is that the link between physical tension and deep emotion in the breathing cycle is a strong one. If, like Norm, you believe you've dealt with past emotions, there's a possibility you've simply packed them away tidily in a box in the corner of the cellar of your mind and they're still imprinted on your breathing—and affecting your life—in the present. For the body, there's no "past" trauma. In some ways, the trauma continues happening now, energetically. Norm and I have a long way to go to fully understand the impact of using breathwork as an intervention for trauma and trapped emotion. But we're excited about what's to come.

WELL-WORN PATHWAYS

We're creatures of habit. Every time you think or feel something, you strengthen the circuitry in your brain, making it increasingly likely that you'll keep thinking or keep feeling that something. You look to confirm these beliefs in your everyday life, even if you're not consciously aware of doing this and even if it makes you feel terrible.

The more time that passes, the deeper these habits are entrenched as the pathways in the brain become well-worn. We can create new pathways to go down, but our brain will continue to try to take the line of least resistance and default to those original pathways it knows best, assuming we do not try to create new patterns with intention. This is another reason why setting your intention at the beginning of this book was so important (Exercise 1). At this stage in our journey it may even be worth revisiting these intention questions to see whether you need to change them in any way.

As you've learned, the habits that you've had for many years are the most challenging to change. These habits are embedded deep in the brain, and by the time you hit 25 years old, you've got so many preexisting pathways that your brain relies on that it's very hard to change them. Sometimes, you've walked the same path so many times that you've created a deep trench—so deep that you can't even see over it to another route.

When this happens, you need something more radical—a transformative intervention, something to help you clamber out of the trench so you can switch paths quickly.

Our brain will continue to try to take the line of least resistance.

CHANGE THROUGH BREATHWORK

The breathwork practice I first walked into with my mum was one of those radical transformations. The technique was similar to the one I've since used with Norm, Danny, Linda, Jasmine and thousands of other people who've had remarkable transformations through changing their breathing. You may have started to feel some of the benefits, shifts and changes already with your daily Infinity Breathing practice.

Breathwork can help a rapid reconfiguration in the body and mind. As we've discussed, I've found it to be a hugely effective way both to release trauma, trapped emotion and old habits, and to change belief systems, all without the need to necessarily talk them through. But, equally, I've found it enormously helpful in getting to the heart of lifelong issues of anxiety, stress, bad sleep, poor digestion, low energy, lack of self-confidence, by addressing these once and for all. The fantastic thing about breathwork is that, unlike conscious awareness and coaching tools, it doesn't need to "know" what the problem is to begin helping you. Breathwork can therefore be incredibly helpful in clearing the unconscious patterns and conditioning that are reflected in how you breathe day-to-day.

Traditional talking therapy is of course incredibly helpful too, but we sometimes neither have the words to explain how we feel nor the answers to solve what's wrong. Sometimes we even corrupt the process unconsciously because we're so used to being ourselves that even our problems feel comfortable—

however counterintuitive that might sound. But all your problems are mapped out in your breathing. There's no hiding from that. And by working with your breathing as an intervention, you can release the emotional tension you hold, shift the chemistry in your body and rewire the pathways in your brain. Think of it like the ultimate reset, like pressing Ctrl-Alt-Delete on your computer.

The breathing workshop I attended with my mum about two months after Tiff's death began my process of dealing with my grief. It cracked me open, enabling me to feel unconditionally. Practicing something that delivers a powerful but short-term form of relief for stress or pain in the moment, such as the "If in Doubt, Breathe It Out" technique, wouldn't have shifted this grief on its own, as my emotional pain was deeply entangled with my past, my core beliefs, my inability to fully feel and let go. I needed to go deeper beneath the surface. Safely working with conscious energy breathwork techniques gave me solace, comfort and consolation, which in turn enabled all the strong feelings that I was holding in my body to come out and express themselves. I could take that bag of bricks off my back. I began to feel lighter, and that gave me the ability to move on. You can do this too.

SOMETHING YOU'VE LOST

The reason I needed a radical intervention was because I was going through grief, a highly complex emotion and an extreme form of social pain. Matthew Lieberman, scientist and author of *Social: Why Our Brains Are Wired to Connect*, says that human beings are hardwired to feel a deep need for social connection because it's key to our survival. A baby couldn't survive long without a strong bond with its parents or caregivers, who provide food and shelter. Anything that risks breaking our social bonds with others is traumatic. It explains day-to-day expressions like "She broke my heart" or "He hurt my feelings." But the real loss of a social connection through grief can be extremely hard to overcome. It is usually accompanied with constricted breathing in your secondary breathing muscles in the midsection and chest. Remember from our previous chapter how the flow of breath around your emotional "heart center" can freeze to protect you from the complex emotion of grief?

Grief can be caused by a wide range of things—it's a severed social connection that doesn't only have to occur through death. You can feel it after a big upheaval, a divorce or breakup, or when you lose your home or job. You can also feel it if you immigrate to a new country, change schools, lose a sense of hope or safety, or lose touch with who you are. Many top sports people experience this type of deep emotional pain inside after losing a big match or even getting silver rather than gold.

Sometimes it's hard to explain to ourselves exactly *what* we've lost. We can lose our sense of self, our sense of purpose, our simple sense of security if we feel the world around us is changing fast and we cannot keep up. When something "goes" that we felt was important to us, all of a sudden our life can feel unstable and unpredictable. And when our future is ineradicably altered, we feel loss and grief for the life we wished we could have led but no longer can—our dream of being a footballer is dashed by a bad knee injury or the trip we always wanted to take is canceled because of travel restrictions. We can also lose our belief in something or someone we once had complete trust in.

And sometimes with these types of loss, we don't feel so much that we're the ones who have lost something, but that it has been taken away *from* us. We can lose a sense of belonging when the world that we know changes radically around us and we feel like we've lost our roots. Refugees fleeing hardship may feel grief at leaving their home behind, even if that home has been flattened by bombs. Directly or indirectly, these are all lost social connections. The pain of loss and grief we feel is very real and something we'll all experience at times in our life.

Breathwork can help you cope with uncertainty, and help you become more comfortable with the change through loss. This is because, unlike loss, it's something we *can* control. We can slow it down to calm our mind, and use it to connect to and release our emotions and express our feelings. When you take control of your breath through uncertainty, you bring a sense of stability to your thoughts and your emotions.

Uncertainty is interesting because it can be a source of pain and something that results from pain. It can make us anxious, and encourage us to try to predict the future. And because we naturally have a negative bias, we may catastrophize and enter a negative spiral. The dedicated athlete suffers a serious injury and wonders if he or she will ever again be able to compete in the event they love. The loving husband loses his wife and doesn't know whether he'll be able to cope without her by his side. Uncertainty makes us feel that we don't have any control over our lives, and that can make us feel powerless and vulnerable to the world around us.

Many of us find this hard to deal with. We like to think that we have control over our lives. We try our hardest to grip on to control, and we struggle when we realize that this is no longer the case. But let me let you in on a secret: we never really have full control over our lives, and it's OK to let go of the need to control.

The idea of letting go of control may sound scary; especially if you are the controller archetype. But we need to learn to let go of the things we can't control, like the world around us, other people's actions, our past, our fears. Letting go of control of what we can't control gives us a feeling of more control, and a sense of freedom. As we accept, forgive and trust, we don't reduce the uncertainty, but we feel less of it. We feel less emotional pain, and that makes us feel we have more control over ourselves and our life. So although I have been talking a lot about control, and teaching you how to take control of your breath to take control of how you think and feel, I also want you to know that you can let go of control to free you from your past.

Complex emotions like grief and loss, or even regret and jealousy, are individual, variable and long-lasting. Your unconscious mind does its very best to try to prevent and protect you from feeling the pain of complex emotions. So your breathing becomes restricted—and you hold on to your past. Symptoms of dysfunctional breathing that make up all the breathing archetypes can come about this way.

The more complex the emotions, or the more entangled with your belief systems they are, the more "stuck" or controlled your breathing can become. The longer these feelings persist, or the breath-holding pattern they have created persists, the more likely you are to feel stuck in the drama of emotional behavior in your life. The bricks and boulders become heavier in the bag that you lug around.

Part of the problem is our inability to express ourselves in the modern world. With the exercises in this book, we have already been learning to get better at feeling by using our breath to access the emotional flow in the body and release it. It absolutely isn't always appropriate to burst into tears and continue like nothing has happened, but bottling your feelings up isn't good for you either. Breathwork practice provides you with a safe, controlled space to let go of that need to control yourself and act "properly." It gives you permission to release the trauma, the stress, the strain, providing you with a safe place to go deep into your psyche to receive wisdom and clarity, which we're often too busy to access.

Your breath is the drumbeat of your life.

EXERCISE 29

INFINITY BREATHING AND LETTING GO

We've talked about control, and how it's important to recognize when something isn't in our control. Up until this point, your Infinity Breathing practices may have been feeling quite controlled. You've been controlling your inhalation, breathing in for three, then controlling your exhalation, breathing out for three. You've been "doing" as you breathe in, "doing" as you breathe out. However, your Infinity Breathing practice is also a really effective intervention that can help you create balance between control and letting go of control. Through it, you can take control and move forward by doing something—breathing in—while releasing and surrendering by just "being"—when you breathe out. It can help you let go, process your complex emotions, heal from your past, drop your bag and empty it out.

The basis of this practice is the same as Exercise 28—Infinity Breathing with Sound and Affirmations.

- Find a comfortable spot, seated or lying down.
- Set a timer for 10 minutes, or have some music ready that will last about that long.
- Allow yourself time to anchor into your body.
- Notice your body, let it soften. Let go of any tension in your face, your jaw, your neck.
- Notice your mind, be aware of your thoughts, judgments and opinions.

- Now feel your heartbeat in your chest and say your affirmations, your three statements.
- Start your Infinity Breathing.
- This time, instead of counting, I just want you to feel.
- Breathe in through your nose, feeling your belly rise and maybe a light, secondary expansion in your midsection, then to your chest. Like a wave flowing up.
- Without a pause, breathe out through your nose, but rather than controlling your breath out, just release it. Completely let go. Do nothing, just be. Let your whole body relax and surrender. Your diaphragm will move back by itself, like a rubber band will after you've stretched it.
- Without a pause, breathe in again, let your belly rise first—open and expand.
- Breathe out. Relax and let go.
- Continue with this Infinity Breath flow.
- Your in-breath is your "doing" state. You are staying "yes" to life. *I'm here, and I choose to be here and move forward.*
- Your out-breath is your "being" state. *I trust and let go.*
- You may wish to say that every out-breath. *I trust and let go* or *It's safe to let go.* To help you surrender deeper with every out-breath.
- It's here at the crossover point in your infinity practice of "doing" as you breathe in and "being" as you breathe out where your bag starts to empty.

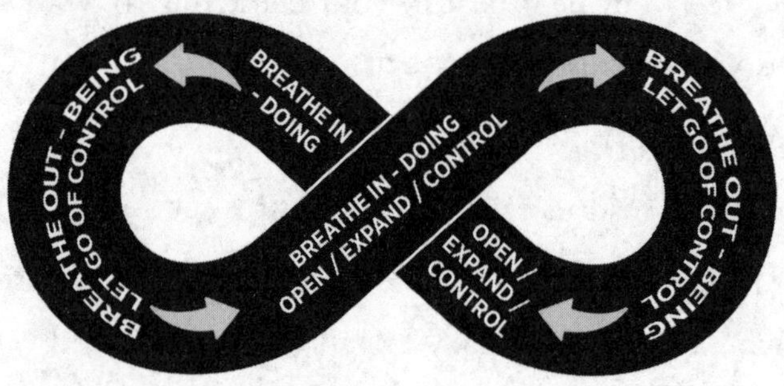

- Whenever you feel the desire, drum your hands on your knees while making a humming sound. You can do this for three rounds before returning back to the Infinity Breathing practice. Remember, if any emotion surfaces, allow yourself to feel it. If you feel uncomfortable, take a break.
- When the 10-minute timer is up, settle and anchor back into your body. Take some slow, deep, relaxed breaths.
- Feel your heartbeat again, maybe even place your hands on your heart.
- Feel the appreciation for your heart beating, for the life and vitality in your body.
- Feel gratitude for all the things that make you feel safe and loved—the people in your life, the resources and opportunities you have available to you. Sometimes this brings a flow of emotion and that's OK—let it flow.
- Come back to your heartbeat and now repeat your affirmations, whatever they are: *I am strong, I am loving, I am supported. I am proud. I choose to be me.*

- And don't just say them in your mind—really feel them in your body.
- Before coming back into your space.

A VALUABLE TEACHER

Complex emotion, despite everything, can be an amazing teacher. We don't always like our teacher and we don't always agree with what we're being taught. Let's take the terrible pain of grief as an example—whether the loss of a loved one, a job or a dream. It imparts knowledge to us whether we like it or not, and that knowledge is valuable. Like almost nothing else, it teaches us that the things we love and value most in the world, even something we barely notice, like living pain-free, can be snatched away from us at a moment's notice. And when we connect with this, it can make us extremely grateful for what we have.

It doesn't just mean that we don't take things for granted, but that we're more likely to seek out connections and try to live our lives meaningfully. We can seek out things of real value and purpose rather than material items. After all, it's almost a cliché that at the end of life, people tend to regret not having spent more time with friends and family. Far fewer people say, "You know what? I wish I'd had more pairs of shoes."

Loss also reminds us that we're all equal. We all lose things, and in the end, we all lose our lives. Death is the ultimate leveler. At the end of the day, each and every one of us departs from this world as we know it, on to a place unknown. I'd like to think we all end up as pure consciousness at a big party in the sky, but there's no concrete evidence to suggest anything lies beyond this life, and even if it does, we can't take anything material with us. So let's use death as an example to help us live more fully now. We don't have to have lost someone, or to be nearing the end of our life, to think about death. We can all benefit at any time by confronting death, as a motivation to live better, more purposeful and meaningful lives. And though it sounds morbid, the opposite is true. Death connects you to living life fully.

One thing I learned from my experience of grief is that, in a way, it becomes part of who we are. You can use your breath to let go of grief's hold on you and clear the debilitating pain. But you may have little twinges of pain that keep you connected to that person or thing you've lost. This might sound strange, since we've spoken about how we can process grief and move on. But we move on *with* our grief.

What I've learned through my experiences, and through helping others to overcome the challenges they face in their lives, is that we're all much, much stronger than we think—by managing what we *can* control, and by letting go of what we *can't* control.

Let's use death as an example to help us live more fully now.

EXERCISE 30
CONFRONT DEATH AND LIVE BETTER

If today was your last day:

- Who would you spend it with?
- What dreams would be lost?
- Who do you want to forgive?
- What would you miss?
- Would you do what you're currently planning to do today?

We can't always do exactly what we want to do every day, or we'd never make progress in our projects. But if you answer "no" to the last question too many days in a row, you may recognize that something in your life needs to change. Confronting death in this way can help keep you aligned with your intentions and what you really want in life.

DON'T BEWARE THE GRIM REAPER

In the West we've become scared of death. We do our best to avoid it and try to prolong our time on earth at all costs, even if that means deep misery for everyone concerned. Many would rather have a wretched quality of life than let it end. But time is just relative. It's the quality of experiences and the life that you live that are important, not the length of time you're alive. When life seemed more precarious, people had a more intimate relationship with death. The ancient Greek philosophers meditated on death and its meaning, and many major religions invoke death regularly, as a reminder that life is fleeting and that we should not get too carried away with earthly pleasures, material satisfaction and external achievements. Some Buddhists practice *maranasati*—mindfulness of death—a reminder that death can strike at any time. In Mexican culture, death is both mourned and celebrated with humor and joy in the Día de los Muertos—the Day of the Dead—which honors those who have passed.

★ ★ ★

Well done for getting here; our deeper-work section can make you confront some deep truths about yourself. In this section we have explored how emotions work, how they get trapped, and how you can shift, process and integrate them with your breathing.

We've considered how the experiences we have over the course of our lives can have a huge impact on our mental well-being, and how even though our trauma is subjective and tangled with our beliefs (remember the ketchup bottle?), it can really affect how we think and feel and behave on a daily basis.

We've discussed how doing and thinking the same things over many years can strengthen connections in the brain, and how a more radical intervention through breathing can help us to finally break these connections and the habits they're associated with, and move forward. You've also met my good friend and mentor Dr. Norm.

From this point forward, I would like you to continue with your 40-Day Infinity Breathing practice introduced in this section, as well as any of the practices and exercises from Part 1.

It's now time to move on to the third and final part of the book, where I am going to share how to optimize your breathing so you can flourish at work, sports and other parts of your life. If you don't think you've done enough deeper work yet to move on to optimizing your breathing, don't worry: you can keep emptying your bag even as we move forward.

PART 3: OPTIMIZE

8

FIND FLOW, FOCUS AND ENERGY

ACHIEVING AN OPTIMAL STATE

So far, you've explored how, by becoming aware of your breathing and making an intention to change it, you can begin down the path to a healthier, happier life. From shutting your mouth (it had better still be shut!) to slowing things down (remember Yogananda?), releasing some of the tension in your diaphragm, you've got some easy, accessible, yet powerful ways to take control of your thoughts and feelings day-to-day. But we've also considered how experiences you may have had 5, 10 or 15 years ago may be affecting you today and how, through a more radical intervention, including your 40 days of Infinity Breathing (don't stop now), you can break patterns, unlearn unhealthy habits, start to empty out your bag of bricks and rewire a new way of thinking, being and behaving.

But I know what you're thinking: What if I don't want *less* of something but *more* of something? What if I want focus and flow, greater creativity, more self-confidence before a big speech? You're not alone. Let's face it: we *all* wish that whether we're heading into the office, holed up in our work space at home or striding out onto the sports field, we could be sure that we'll perform at our very best. Even after all our deeper work, life can throw curveballs. Maybe the baby kept you up all night, or you feel weighed down by that late-night pizza you had, or you were so anxious about the day ahead that you tossed and turned all night. Perhaps the ultimate distraction—your phone—has stolen your attention. Daily high performance—effortless flow, Jedi focus, constant energy, boundless creativity—seems like an impossibility, and the people who manage to do it appear superhuman. Achieving this optimal state, however, is easier than you might imagine.

FIND YOUR FLOW

Ever had one of those days when you feel like nothing can stop you? You're bringing your A-game every step of the way. *Pow, pow, pow.* You're batting away "to-dos" like Serena Williams, gliding effortlessly from one task to the next. Your energy's up, yet you're peaceful and calm. Your breathing is steady, time no longer exists, and there's no self-consciousness or mind-wandering. You're right here, right now, fully engrossed in what you're doing.

This is what we call the "flow state," the sweet spot between stress and rest. Athletes call it being "in the zone." It's the optimal physiological and psychological environment for peak performance. It's the state when a musician's jamming and everyone is dancing, when the entrepreneur's captivating a room of investors with a pitch for a game-changing idea, when the Olympic athlete is flying down the home straight of a record-breaking run. We all want to feel this way. And we want it more than ever on those days when we've got no choice but to perform well: the day of the job interview, the best man's speech, the big pitch, race day.

Chances are that you've been in this state before. Ever noticed how, when the conditions are just right (like when a holiday's around the corner and you *really* need to get some things done before you go), you find you can get a week's worth of work done in a couple of hours?

That's flow.

A VIRTUOUS CYCLE

The concept of flow was first described in the 1970s by Mihaly Csikszentmihalyi (a tongue-twister of a name, pronounced "me high, cheeks send me high"). He's a Hungarian-American psychologist and one of the pioneers in the field of human happiness. He studied thousands of high achievers, and found that if a task was just difficult enough to challenge someone's skills, they would both want to do it and feel satisfied completing it. By successfully completing the task, they would become more skilled at doing it, and would want to do a harder task. Flow therefore creates a virtuous cycle both of challenge and accomplishment.[46]

In other words, when we complete a task that's just beyond our level of ability, we feel hugely satisfied and we improve. We're then compelled to go on and try to complete more challenging tasks so that we keep experiencing this positive feeling. Flow and high achievement often go together, because as we improve, we need to keep testing ourselves to keep accessing it. This means that we get better and better. A 10-year study showed that people in flow states were five times more productive.[47] Think how much you could get done in a day, or how much spare time you could have to relax if you found your flow when you wanted it!

So how do we get into this state in the first place? As Csikszentmihalyi says, we begin by balancing our ability with the difficulty of the task. That means ensuring that it's not so easy that we get bored, and not so difficult that we're put off or give up. But there's another powerful way to access this state. And that's breathing.

By practicing a breathing pattern that promotes a balance between your stress response and your rest response, your Heart Rate Variability increases, and you send a coherent heart rhythm to your brain. This helps align a number of systems in the body to this rhythm. We explored a version of this with our Magic Ratio Breathing.

Flow and high achievement often go together.

A MATTER OF LIFE AND DEATH

When turning stress to calm in Chapter 4, we learned that breathing is quite binary. By breathing in more, you flick the "on" switch, triggering the sympathetic response and increasing your heart rate. On the other hand, slow breathing and taking longer breaths out flicks the "off" switch, triggering your parasympathetic state and lowering your heart rate. Because being in a flow state requires us to balance our stress and rest responses and have coherent heart rhythms, we need to balance our in-breaths and out-breaths to access it.

You don't just have to take this from me. The US Navy SEALs, considered among the best and most highly trained special operations forces in the world, do just this to quickly get into a flow state before entering a dangerous situation. They don't want to be stressed, but nor do they want to be completely chilled out. These elite military units need to be hyperfocused on the task at hand and resist getting distracted by the myriad things they face in the line of duty. It's literally a matter of life and death. So if it's good enough for them, I think it's good enough for us.

EXERCISE 31
BOX BREATHING

The technique the SEALs use is called box breathing, and it's one of my go-tos in the intervals between meetings or client sessions to optimize my mind. Even when I'm walking from A to B, or anytime I need to mentally rest and find my flow, I turn to this technique. It's also a great way to equalize the oxygen and carbon dioxide in your body.

Box breathing gets its name from having four sides. Let's give it a go:

- Breathe in through your nose for four, using your diaphragm, feeling your belly rise.
- Hold for four, keeping calm and still.
- Breathe out of your nose for four, steady and controlled.
- Hold your breath for four. Be soft and gentle with yourself. (No clamping or tension: make your breath holds soft and delicate.)
- Repeat.

Just one round will have a positive impact on your mind and body, but a flow state doesn't happen instantly. It takes time for the heart, body and mind to synchronize and entrain to this breathing rhythm. Some research shows that it can take up to 15 minutes to access this state,[48] but I've found that, with practice, just four minutes of box breathing can get you close. Try to commit to this for the next four minutes if you can. You can even try it as you read on too.

If you find that over those four minutes you're feeling too much air hunger (you'll know when this is because you'll have an uncontrollable desire to breathe), start with a pattern of 3-3-3-3 and increase it over time. It can be enjoyed in stillness while sitting with your eyes closed or when you're walking, matching the counts with your footsteps—four steps breathing in, four steps hold, four steps breathing out and four steps hold. I like to add this way of breathing on my daily commute as I walk to the office, but you can practice it when you're walking to the shop, moving from room to room or out walking your dog—anytime you get the chance.

When box breathing is practiced in a group, something special happens—everyone's heart rhythms synchronize.

Bodies and minds become one as the group's energy entrains to reach a state of collective flow. You can imagine how powerful it is for a special operations force, like the Navy SEALs, to think with one collective brain, one looking left, one looking right. I've found box breathing to be particularly useful when working with business teams, such as when a group needs to think and feel more collectively and coherently before a pitch, project or task.

JEDI FOCUS

You can't talk about flow without talking about focus. But, contrary to popular belief, they're not the same thing. Without focus, you cannot be in a flow state. But without a flow state, you *can* focus. You may feel like you've got too much focus on the negative and wish you could focus less, maybe after that time you said something you regret and are unable to think of anything else for days or if your mind is obsessively worrying about something in the future that you've no control over. Well, in a *purely* focused state—when you're only focused on the moment itself, and not pursuing thoughts—the notion of past and future doesn't exist as you've become fully present, immersed in the moment. When you're really focused in the truest sense, not just directing attention toward something but wholly engaged, your mind cannot wander and rehash old mistakes or ruminate about the past, nor can it worry about the future. You just focus on what matters now.

All of us wish we could be just a bit more focused on the things that matter to us. And when we want to get focused, the first thing we have to do is pick one task only—I hate to break it to you, but multitasking is a myth. When we try to do many things at once, we overwhelm ourselves, our thoughts swirl and we find it hard to think clearly. A conflict arises in our mind as we don't know what to do first. We're chasing different things at the same time, we're thinking chaotically and we soon tire ourselves out. And though we might tell ourselves that we're great at spinning plates, the truth is we're not. Even in the very best-case scenario, we're making progress much more slowly than we would if we did things one at a time. According to computer scientist and psychologist Gerald Weinberg, when you're working on multiple tasks at the same time, you deplete your productivity by up to 80 percent. Think about it this way: if you're focusing on one task, you're giving 100 percent of your productive time to it. If you're doing two things at once, you're giving 40 percent of *your productive time* to each, and losing 20 percent to context switching. Once you increase this to three tasks, you are only focusing for 20 percent of the time on each and losing 40 percent to context switching. And so on.[49] So not only is our attention limited, but when we multitask we're effectively distracting ourselves by choice and we can guarantee we won't reach a flow state. And it isn't like we need any help with that. Our world is full of distractions. It seems sometimes as if it were designed specifically to stop us from focusing. Our phone fires notifications in our direction a gazillion times a day, dozens of tabs on our internet browser compete for our attention, there are so many shows on our TV we can't even stick with one for a whole episode (and we're probably using our phone while we're watching anyway). The workplace is full of distractions too. Every time we're distracted, we lose our focus and either stop ourselves from getting into the zone or yank ourselves out of our flow state. The clock resets and we have to get reabsorbed all over again into whatever it was we were doing.

Nowadays we don't always have the luxury of eliminating distractions. In a busy office, there's noise everywhere. The nature of some jobs, like customer-service roles, is to react to things as they happen. Stay-at-home parents and home workers often have to try to get through their to-do list while juggling the needs of their kids. And top athletes have to focus on the task at hand while surrounded by thousands of screaming fans. As for those Navy SEALs, getting shot at comes pretty high on the list of distractions.

LAMA IN THE LAB

Researchers interested in the startle reflex, the intensity of which predicts the level of negative emotion that someone has, tested Buddhist monk Matthieu Ricard, who under the pseudonym Lama Oser had racked up tens of thousands of hours of meditation practice. They counted down and played him a gunshot sound at the very top of the human sound threshold. Although a study in the 1940s concluded that it was impossible to prevent the startle reflex (even police marksmen, who fire guns routinely, are unable to keep themselves from being startled), Lama Oser was able to suppress his response completely. He later said: "If you can remain properly in this meditation state, the bang seems neutral, like a bird crossing the sky."[50] Although we can't all become hermit monks like Lama Oser, we can lessen the extent to which we're distracted day-to-day by using our breathing.

If we cannot eliminate distractions, we need to minimize them and become less reactive to them.

WHAT IS A DISTRACTION?

What is a distraction? We all know what distracts us: that hilarious video sent by our mate, a new post on our socials. Evolutionarily speaking, being distracted by surprising sounds was important in ensuring that we were attentive to our environment in case a threat or an opportunity appeared. But a distraction is just something that takes your attention away from what you're doing or thinking. What distracts one person might not distract another.

Some people, like our Navy SEALs or a good waiter in a busy restaurant, might not register certain stimuli as "distractions" to the same intensity that others do, either because they're so used to them or because they have highly developed powers of focus. Of course, there are also reasons we might actually want to be distracted—for example, parents will want to be able to hear their child when they're in the other room. But most of us are easily distracted and wish we were not, and the key thing to remember is that if we cannot eliminate distractions, we need to minimize them and become less reactive to them, like Lama Oser. The good news is that the better we get at focusing, the less we'll register distractions that throw us off and disrupt our flow, and the more we practice, it will then become harder and harder for distractions to jolt us out of what we're doing.

How do we do this? Well, the first step is to minimize distractions. This could mean putting your phone on airplane mode. That helps me a lot. Better still, you could leave it in the other room, switch off any notifications and let people (friends/family/colleagues) know not to bug you (that's an important one). Now, of course, the distraction might actually be your own mind—those racing thoughts and ideas. As I mentioned before, some of our focus may have tangled itself in the future or past, creating worry or a loop of overthinking.

THE OVERTHINKER

Hands up if you worry about things that are out of your control—your environment, the outcomes in situations, other people's actions, even the weather? The good ol' Dalai Lama says:

> *If a problem is fixable, if a situation is such that you can do something about it, then there's no need to worry. If it's not fixable, then there's no help in worrying. There's no benefit in worrying whatsoever.*

And yes, I get it, Your Holiness D.L. But saying, "Don't worry about it" while you're in a vicious cycle of "Worry, try to gain control, fail, worry more, repeat" is really tough.

Here are a couple of tips to help you take control of your overthinking mind so that you can approach things differently:

Tip 1: Identify what's in your control and what isn't

- If you're faced with a problem and are suffering the discomfort of overthinking, ask yourself whether it's a problem you can solve. Maybe you need to change how you act and feel about the problem?
- If it's in your control, tackle it.
- If not, is there something you can do to inspire action?

- For example, you can't force your team to be productive, but you *can* give them the tools and support they need to succeed. You can't force someone to change their diet, but you *can* share with them an inspiring healthy cookbook.
- If your mind is still racing, focus your mind on the present with some rounds of Jedi breathing, with or without your hands (Exercise 32).

Tip 2: Schedule a slot of "worry time"

Sounds mad, right? But there are studies to back its usefulness.[51]

- Pick a 15-minute slot each day in which to worry (just not before bedtime!).
- If you worry outside of your time slot, remind yourself it's not worry time.
- During your 15 minutes of worry time, write down everything you're worried about that's outside your control.
- When time is up, carry on with your day. You will soon begin to contain your worry in 15 minutes, which is better than worrying 24/7![52]

EXERCISE 32
JEDI BREATHING

Breathing can make all the difference when you're trying to focus your mind and break the loop of overthinking. One technique I use is something I like to call "Jedi breathing." It's our alternate-nostril breathing technique from Chapter 3 with a twist. So let's start with a round of this. It has balanced in-breaths and out-breaths, like box breathing, but with the difference that we switch nostrils, which forces us to concentrate even more.

- Make a peace sign with your right hand using your thumb and ring finger.
- Close your right nostril with your right thumb, and breathe in through your left nostril steadily for a count of four.
- Pause as you close your left nostril with your right ring finger and open your right nostril.
- Breathe out calmly through your right nostril for a count of four, pausing briefly at the end of your out-breath.
- Breathe in through your right nostril steadily for a count of four.
- Pause as you close your right nostril with your right thumb, and open your left nostril.
- Breathe out calmly through your left nostril for a count of four.

- Now you have the hang of the alternate-nostril breathing movement, it's time to go into Jedi breathing mode. I want you to try the same technique, but *without your hands.*
- Breathe in through your left nostril for four.
- Breathe out through your right nostril for four.
- Stay with me! Really focus!
- Now breathe in your right nostril for four.
- And breathe out through your left nostril for four.
- Repeat.

I get it. It's really hard. And it requires a lot of focus to bring your awareness solely to the flow of air in each nostril. But let me ask you this: When you were practicing this technique, what else were you thinking about?

This technique is so challenging that it forces you to think solely about your breathing. So your mind's awareness, which is often vast like the sky, goes to that one thing. You're breathing with such intent that the mind moves into a highly focused state, your thoughts being channeled to a singular point like a laser beam. You're no longer thinking about the past or worried about the future. You're focusing on the present. Your breath, and the air flowing in and out of each nostril.

This is an especially effective thing to do before you need to study, revise, or work on a project that requires your full attention. Or when you're feeling particularly distracted by your surroundings or your mind. Your attention is taken away completely from everything that's going on around you so long as you're focusing on your breathing. The noise of the day quietens, and you're now primed to focus intently on whatever you need to do next.

Now you've practiced box breathing to find your flow and Jedi breathing to improve your focus, and you've eliminated as many distractions as possible. But your brain is a greedy organ, devouring up to a quarter of your energy. You need fuel to sustain your concentration over time, and if you had to stay up late, or you had broken sleep because of your kids or your neighbors' house party or jet lag, you might just want something to give you that boost to get you through your day.

Box breathing may help to keep you balanced, but sometimes you might actually need to flip the "on" switch and create a sympathetic stress response. Let me tell you why.

STRESS IS GOOD

While the benefits of the parasympathetic system appear self-evident to most people, those of the sympathetic system are less obvious. Why on earth would I want to force myself into a state of stress, you ask? Well, stress is *good*, at least sometimes, although it's rarely invited to the party and is often deemed the bad guy. But there's good stress as well as bad stress.

You've already learned that stress is how your body saves your life (remember our grizzly bear?); in addition to this, there's such a thing as good stress, known as *eustress*, which has a short-term effect and feels exciting. It can make you stronger, faster, more energetic, more productive. It can motivate you, and get you ready to act, react and perform when it counts—like before an important meeting, a wedding, moving house or getting a promotion.

Take it from Dr. Kerry Ressler, professor of psychiatry at Harvard Medical School. She says that "a life without stress is not only impossible, but also would likely be pretty uninteresting—in fact, a certain degree of stress is helpful for growth."[53]

Maybe you need a bit of get-up-and-go or something to help you beat that afternoon slump. Stimulating the sympathetic response should be done sparingly, like drinking coffee; we all know how off we can feel when we've just knocked back our sixth espresso.

Breathing trumps caffeine because it doesn't interfere with the adenosine receptors in the brain, which we learned about in Chapter 4.

EXERCISE 33
BELLY-CHEST EXHALE

This exercise is perfect when you need a bit of an energy boost. It's something I've taught DJs who want a healthy way to get pumped up before a set so they're on the same level as the crowd. I also find it a useful exercise to open up the sections of the breath, which is helpful if you're still finding that your chest is collapsed and you're not flowing in your emotional center. A couple of rounds of this will also help ramp up the flow of energy at the start of your daily Infinity Breathing practice.

- Take a half breath through your nose into your belly.
- Take another half breath through your nose into your chest.
- Take a full breath out through your nose.
- Repeat three rounds of 16 reps.

The action is driven by your diaphragm and intercostals; if you are just sniffing hard with your nose, you are likely to feel a bit dizzy or light-headed, so take a break and practice at a pace that suits to get the movement right. Doing it slowly can help open up restrictive patterns in your chest, while practicing at a more dynamic pace will really get your energy flowing.

Music also helps with this exercise. Press play on some energizing tunes and turn up the volume!

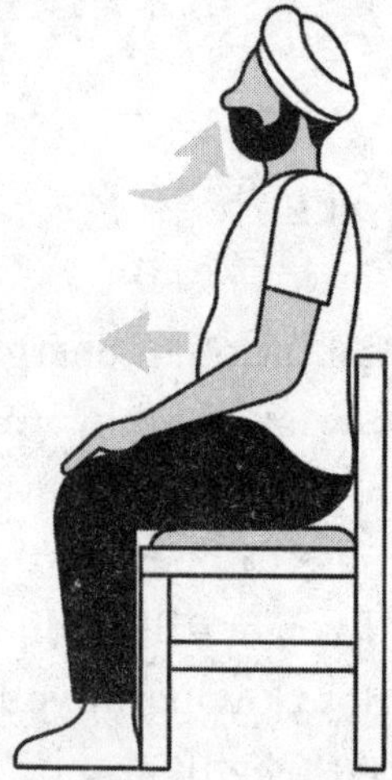

Take a half breath through your nose into your belly.

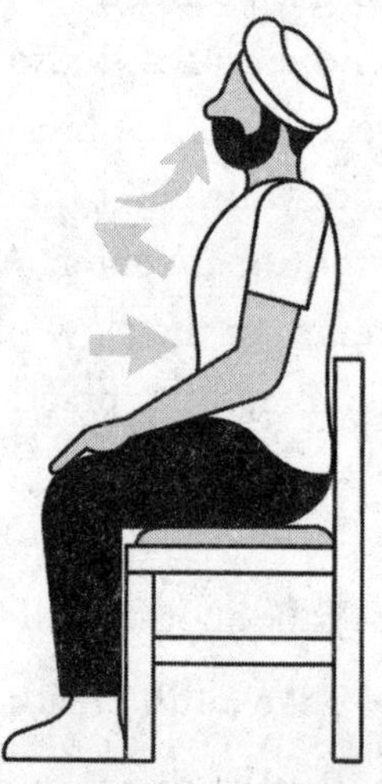

Take a half breath through your nose into your chest.

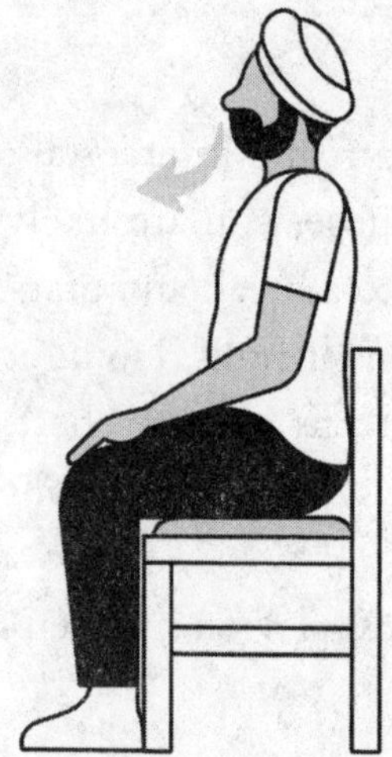

Take a full breath out through your nose.

EXERCISE 34
BREATH PUMP

If you need another way to ooommph your energy, the following exercise is good. It differs from the Belly-Chest Exhale exercise by taking your energy levels up a notch further. Only use it when you need a sudden burst of energy. (I like to use it to combat jet lag.)

You're triggering your sympathetic drive in this exercise, so you don't want to do it all day, much like you wouldn't drink coffee all day. If you feel dizzy, giddy or light-headed, stop. If you're menstruating or pregnant, please miss this one out; it's important during these times that you nurture and rest your body, and this exercise does not do that!

The exercise is powered from the navel point (belly button), as it's pumped in and out on each exhalation and inhalation respectively.

- Sit up tall with a straight spine. Your chest will remain relaxed and slightly lifted.
- First, to get the navel movement, I'd like you to cough. Can you feel your belly button move back toward your spine? Now close your mouth and mimic this movement in the following way.
- Breathe in quickly through your nose, engaging your diaphragm; belly moves out.

- Breathe out through the nose quickly, engaging your core; your navel pumps back toward the spine.
- Your inhalation and your exhalation should be of equal duration.

Once you've nailed the movement…let's really get things going, by moving your arms too.

- Breathe in through your nose quickly; feel your belly rise while lifting your hands up in the air above your head.
- Breathe out through the nose quickly, engaging your core, and pump your navel back toward the spine, while bringing your elbows to your sides.
- Repeat by breathing in and out of the nostrils, pumping your navel while lifting your arms up and down.

Breathe in quickly through your nose, your belly moves out.

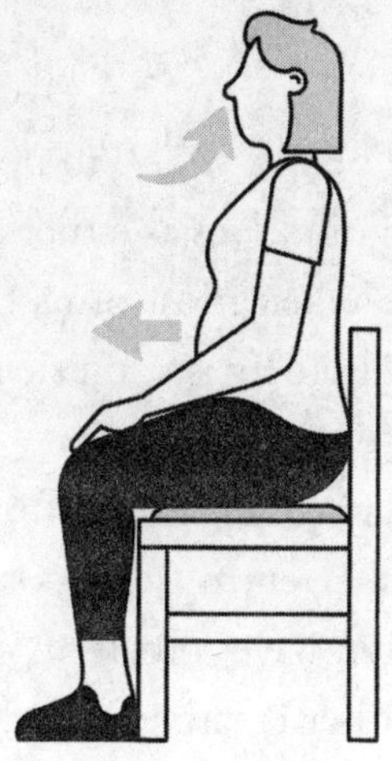

Breathe out quickly through your nose, your belly moves in.

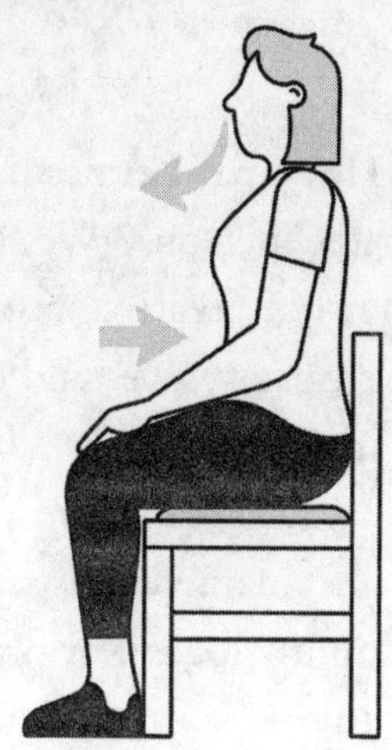

Breathe in quickly through your nose, your belly rises as your arms move above your head.

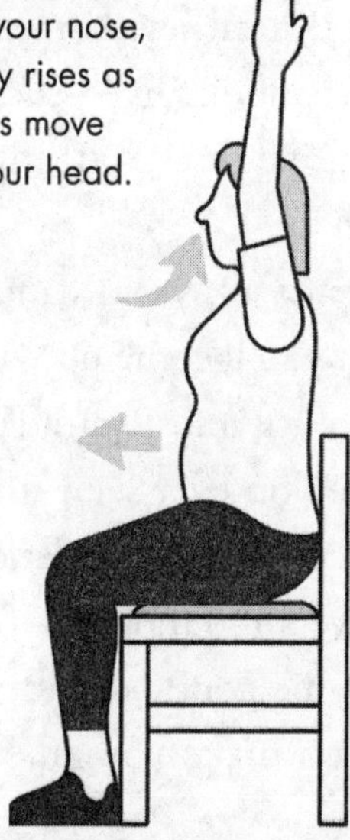

Breathe out quickly through your nose, your belly moves back while your elbows are brought down to your sides.

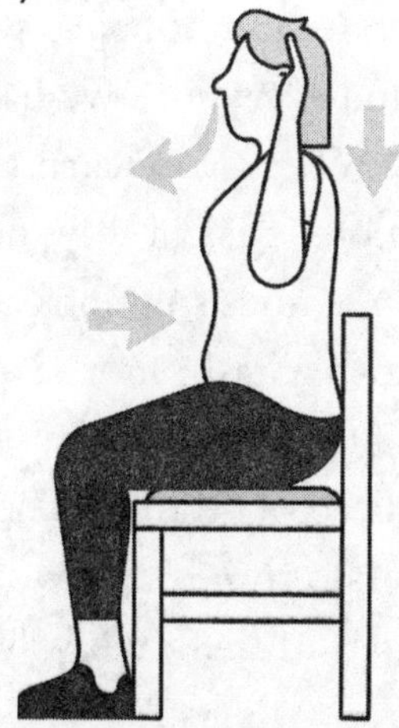

BOOST YOUR CREATIVITY

How many times have you heard people say, “I’m not a creative person”? We like to think of ourselves as either creative or not creative, probably because when most of us think of creativity, we tend to relate it solely to artists, musicians and innovators.

But creativity is more than that; it’s all around us and is a fundamental aspect of our lives. Even in fields considered highly logical or practical, we often have to call on our creativity. It’s a myth that jobs are either creative or functional, although of course many skew one way or the other. Everything we think and do is creative. We are creative so naturally that we don’t even realize it, especially when our day is filled with repetitive tasks. We tend to tune out and go on autopilot just to get things done. But rather than assigning creativity simply to the artistic parts of our lives, such as our crafts and hobbies, we can be creative in everything we do and every idea we have.

When you learn to harness creativity, you’ll find better, more efficient ways of functioning, you’ll come up with more ingenious solutions and even make the most mundane day-to-day tasks more enjoyable. Have you ever seen the videos of the roadside sign spinners? Or policemen who dance while directing traffic? These people have all found new, fun ways to do things that would ordinarily be considered quite boring; in the same way, creativity can make you enjoy whatever it is that you have to do. And the best thing is, the more you use your creative power, the more creative you become.

The British social psychologist and educationalist Graham Wallas identified four distinct stages to the creative process: preparation, incubation, illumination and verification.[54] That can be tricky to remember, so I like to think that creativity is like baking a cake. In Wallas's first stage, you get your ingredients together—you prepare by gathering information and inspiration, making mood boards and trying to learn as much as possible about whatever project or problem you're working on, such as how to make something mundane more enjoyable. You might need time to yourself to really focus on a problem you're trying to solve as you blend and mix the ingredients, but you might just need to get out and be inspired by the world around you. In the preparation phase of creativity, Jedi breathing can be useful, especially if you're prone to distraction.

In the next stage, incubation, you simply put your ingredients in your cerebral oven. You completely defocus and let all the material you gathered up simmer in your brain. This can be hard for those who are a controller archetype, always wanting to be doing something, and often competitive, ambitious and impatient, as the process happens most effectively when the mind is at rest, including while you sleep. I sometimes refer to this as defocused activity, when you stop focusing intently on a task but the ideas and influences you've focused on in the preparation phase are playing in the back of your mind, forming new connections and generating insights. Einstein called this unconscious process "combinatory play." We can enhance this stage by doing something relaxing—going for a stroll, having a bath or shower, taking a nap or playing with our Infinity Breathing practice.

Stage three—illumination—is when we take the cake out of the oven. This is sometimes called the "Eureka!" moment, after Archimedes's moment of insight. But for you, this moment might come during your daily breathwork practice. These epiphanies are moments of pure creative intelligence, and they're rare because our daily thinking is an almost nonstop process. It's only when the stream of consciousness stops scanning the memory that it can quieten for long enough to allow a new electrical impulse of thought to emerge. This fully formed new idea can only finally come to us when the mind is at rest and the endless internal chatter of our stream of consciousness is silenced. The experience of having one of these moments is often hugely satisfying—let the emotions flood your body.

Finally, in the fourth stage—verification—we get to taste our cake and see whether it's been blended with the perfect mix of ingredients. If it's not, we'll start to tweak our idea until it's just right. We bring our ideas into our music, our business, our book, and then build on them. Think of it as like sifting for gold: we need to give it a clean and remove anything inessential so that we're left with a perfect, usable nugget. This may require a little focused Jedi breathing to help add those finishing touches.

The more you use your creative power, the more creative you become.

CHANGING BRAIN WAVES TO HARNESS CREATIVITY

Remember our brain waves in Chapter 6? Some of the greatest minds in history found novel ways to help them formulate their major insights. Both Albert Einstein and Thomas Edison relied on the early, nodding-off stage of sleep—when theta waves predominate—to chew over big ideas. Dmitri Mendeleev dreamed up the periodic table of elements in a deep delta state, while Salvador Dalí learned how to surf between theta and alpha waves to gain new ideas and find creative solutions.

At the other end of the day—early morning—it's possible for individuals to stay in the creative theta state for between 5–15 minutes, making your first moments of consciousness the perfect time for the free flow of ideas. This time can be extremely productive, a period of meaningful and creative mental activity. So it's unfortunate that for most people in the last couple of decades these vital creative moments of our mornings have effectively been hijacked. A survey that studied over 7,000 smartphone users found that 80 percent of the respondents used their phone within 15 minutes of waking.[55] We're allowing our brains' most creative time to be taken over by other people's ideas—their social feeds, pictures, videos and messages, rather than giving ourselves the time to arrive at something new and personal. I'm guilty of jumping on my phone first thing sometimes, but treating yourself to those first 15 minutes, phone-free, for theta-state idea creation, contemplation and journaling will have a profound effect on your mood for the rest of the day.

Deeper delta brain wave states were only ever thought to occur naturally during sleep—and in those fleeting moments just before or after sleep—until a series of experiments between 1970 and 1977 undertaken with famed yogi Swami Rama. These experiments were the first of their kind and shattered many of the assumptions of Western science. In the lab, Swami Rama was able to use his breathing and meditation to change the temperature of different parts of his hand by 10 degrees Fahrenheit, as well as raising his heart rate at will, from 70 to 300 beats per minute. He also voluntarily stopped his heart from beating for 17 seconds. In a further study, Swami Rama managed to change his emitting brain waves to theta and then delta. Previously only thought possible in deep restorative sleep, Swami Rama remained fully conscious of his environment and was able to recall everything that was happening when in these states. He showed that human beings have the potential to move into altered states and take control of these states to help boost creativity, and restore the body and mind.

As adults, we spend most of the day in a beta brain wave frequency. This state is characterized by our "doing mode": active thinking, problem-solving, focusing on tasks and just generally ruminating.

However, through your daily breathwork practice, your brain frequency begins to slow down to match your breathing and your brain emits alpha waves. You become calmer, more introspective, and enter into your "being mode," which in some ways acts as a bridge between the conscious and the subconscious. This state naturally occurs when, for example, you take a relaxing sigh of relief after handing in a big project, and every night before you nod off. If for some reason your brain isn't producing very many alpha waves, it's probable that you're in an overanxious state of mind.

With practice in accessing alpha states, you can learn to slow your brain waves further and experience deeply relaxed yet alert theta frequencies. Theta states are connected with intuition, creative insights and daydreaming, enabling you to access memories, emotions and sensations in the subconscious mind. Theta waves are dominant in that lucid moment just before falling asleep or just when you wake, and are also emitted during deep-focus states of the kind induced by meditation and prayer. These are also the brain waves that are emitted when you enter "defocused" activity, such as those moments of insight in the shower or bath, or even while shaving or brushing your hair—states that come about when a task becomes so automatic you can mentally disengage from it and enter a free-flowing place of ideation.

SHOWER GENIUS

The website Reddit has a subpage called www.reddit.com/r/Showerthoughts, where users post insights they've had while in the shower. Our ability to arrive at brilliant insights while washing ourselves has been studied by cognitive scientist Scott Barry Kaufman, co-author of *Wired to Create*. His study highlighted "the importance of relaxation for creative thinking," as he found that 72 percent of people have creative ideas in the shower. "The relaxing, solitary, and nonjudgmental shower environment may afford creative thinking," he stated, "by allowing the mind to wander freely, and causing people to be more open to their inner stream of consciousness and daydreams."[56]

The vital creative moments of our mornings have effectively been hijacked.

PUBLIC SPEAKING

Like the vast majority of people, I've been terrified of public speaking all my life. At school I'd try to conjure up any excuse I could think of to avoid presentation day: "I've got food poisoning," "I'm locked inside my house," "I've been kidnapped by pirates on the way to school." It was no different at university, even after years of competing in judo in front of crowds of hundreds, and just the same when I was a DJ playing in front of thousands. Hand me a mic and suddenly I'm Silent Stu.

This frustrated me. I could speak one-to-one with anyone, and if you'd asked me I'd have told you I was a confident person. But as soon as I was in front of a group I'd change, and a paralyzing fear would kick in. My hands would shake, my heart would pound, heat would rush into my shoulders, neck and face. I'd blush, my mouth would dry out and my voice would quiver. I wouldn't even recognize my voice. And I'd think, "Who is this?" If something changed, like the lighting, I'd completely lose track of what I was saying and just stand there in silence. "Keep going, Stuart!" went the voice in my head. But most of the time I'd simply freeze.

As far as my unconscious mind and body were concerned, public speaking was like noticing a tiger in the room. I'd moved through fight or flight and was in freeze mode. Then the paranoia would kick in, the negative thoughts, the embarrassment. I know I'm not alone in having had these kinds of experiences. Fear of public speaking, glossophobia, is believed to affect up to 73 percent of the population, according to a report by the National Institute of Mental Health.[57] Even renowned speakers, including Abraham Lincoln and Mahatma Gandhi, suffered from severe anxiety when they had to give a speech publicly. Gandhi would say that "the awful strain of public speaking" prevented him for years from speaking up even at friendly dinner parties.[58] This fear can be caused by a combination of genetic tendencies or environmental, biological and psychological factors. People who fear talking in front of people may have a fear of being shown up or rejected. It might relate to a single bad experience of public speaking—one that was so bad that you can't bear taking the risk of ever having it again.

Even if we think we may never have to speak in public, it's a fear that's important to confront. You might be asked to speak at a best friend's wedding (or be expected to do so at your own). You might find you want to give a eulogy for a loved one that really shows what a fantastic person they were. And if you *can* get comfortable with public speaking, you'll find it benefits your life in ways that you didn't realize it would. You can add more value in your work, boost morale among your teammates, feel self-confident in a group of strangers, even tell better jokes.

Public speaking was like noticing a tiger in the room.

OVERCOME PUBLIC SPEAKING ANXIETY

Nervousness is normal. The best way to overcome public speaking anxiety is to prepare, practice and breathe.

- *Prepare & Practice:* Take time to plan, and go over your notes. Once you have become comfortable with the material, practice, practice, practice. You could even film yourself on your phone, or get a friend to critique your performance.
- *Breathe.* If in Doubt, Breathe It Out. (Exercise 14)

This will turn your stress to calm and remove the tiger from the room…

Once you're more comfortable with public speaking, you might start to enjoy it. Then, and you might want to use different exercises, to get in the zone—maybe box breathing (Exercise 31) to find your flow, or Jedi breathing (Exercise 32) to gain focus. Or one of our energizing exercises (Belly-Chest Exhale, Exercise 33, or Breath Pump, Exercise 34) to get pumped up before your performance.

PAVAROTTI'S IMAGINARY FRIEND

Public speaking is a kind of stage fright or performance anxiety, and some of the biggest names in the world of entertainment have suffered from it. One of the best examples is the story of the great Italian opera singer Luciano Pavarotti. Legend has it that he had to conceptualize his fear of performing as a small child in order to find the confidence to go out on stage. Before every concert, an imaginary little boy called Fear would knock on his door and, hand in hand, they'd go out in front of the crowd together. Whoever we are, we all have to find ways to conquer our fears, and this was Pavarotti's.

EXERCISE 35
WELL SAID

Read this paragraph out loud:

Growth happens when you step outside your comfort zone and start doing the things that you haven't done before. Stop telling yourself you can't, you're not experienced enough or you're not good enough. Practice, prepare and breathe. You've got this.

OK, how did you breathe in between sentences? Nose or mouth? Most people that come through my door mouth breathe in between sentences when they speak and wonder why they feel stressed, anxious and tired. Remember in Chapter 4, I told you to shut your mouth? Speaking is no different. We're never really taught to breathe and speak. If your work demands that you speak a lot, like teaching or sales, or you are someone who often tries to get your point across (like our breath-grabber archetype), it's likely you will have defaulted to a gasping mouth-breathing pattern when speaking. And you know how problematic mouth breathing can be.

You need to learn to breathe in through your nose and talk out your mouth. It's a tricky one to get the hang of, and it will take a bit of practice to find your rhythm. As always, awareness is the first step, and even if you manage to reduce the amount of mouth-breath speaking, that is a good start. When you make the switch to your nose, your speech will naturally slow. This may feel strange at first, but it will in fact make you sound more authoritative, considerate and in control of your words.

Practice makes perfect with this one. Try reading the passage out loud again, but this time breathe in through your nose and talk out your mouth. Repeat.

STATES BECOME TRAITS

There's a saying in neuroscience: "Neurons that fire together, wire together." The more we think something, the more likely we are to think it. And this process is extremely powerful. The LA Lakers, one of the most successful basketball teams in history, were divided into groups to test their powers of visualization. One group practiced free throws and another were asked only to visualize and *think* about practicing free throws. The result? The group that only thought about, but did no physical practice, improved almost as much as the ones taking shots on the court.

This phenomenon applies to emotional states as well. If we're anxious, we're more likely to keep being anxious as the paths in the brain relating to anxiety become more and more well-trodden. And if we practice getting ourselves into positive emotional states, then we're more likely to feel positive. Over time, our states can become traits.

This is why it's so important to keep practicing using our breathing to improve our focus, creativity, and other positive mental and emotional states. The more we do it, the better we get. And this applies to all the exercises contained in this book. It's also why it's so important not to feel discouraged if you find that Navy SEAL–level laser focus doesn't materialize overnight. Practice makes perfect—so keep with it.

Few people know more about the importance of practice than athletes, gym-goers and others looking to improve physical health and fitness. That will be the focus of our next chapter.

9

IMPROVE YOUR FITNESS

AN UNTAPPED RESOURCE

"40 seconds left!" Billy shouted. "Now hold your breath and keep going!"

Billy was my judo coach. He's been involved in every Olympic Games, either as a player or coach, since Barcelona 1992. As part of our training, he used to make us hold our breath while we practiced *uchi-komi*, throwing practice drills. I hated this—it felt like some sort of torture—and I assumed it was just one of his ploys to toughen us up. Billy often had unorthodox ways of training us, and this seemed like another one of his wild ideas. It was only years later, when I was leading a performance breathwork session with some of the national judo team and ran into him, that I decided to ask him why he made us do it.

"The Korean team were doing it, and they were the best in the world at that time," he said with a shrug. "So I added it to your program."

Now, Billy may or may not have realized it at the time, but it turns out there's a lot of science behind the Korean team's approach, and it may be one of the reasons they got to the top. I'll explain how in a moment. But first, let's explore the relationship between breathing and performance.

Whether you're going for Olympic gold, trying to beat your PB or just improving your general fitness so you can play with your kids without feeling exhausted after five minutes, breathwork is hugely useful. Understanding how breathing relates to fitness and performance will not only help you make progress at your given sport, but will also improve the efficiency of your day-to-day breathing and enhance your health.

Have a think now. What stops you from running that extra mile or doing one more round in your workout or fitness regime? What makes you pause for a break when you have to take the stairs that day the lift is out of order?

Well, what happens usually is you "gas out." An overwhelming feeling of breathlessness forces you to stop, or your muscles get fatigued and start to ache unbearably due to the buildup of lactic acid. Sometimes it's both of the above.

Because oxygen is the fuel for your muscles, the more activity you do, the more oxygen you need. Oxygen gives your muscles the energy they need to sustain increased activity. And while this is happening, you also create more carbon dioxide. You'll remember from Chapter 4 that the desire to breathe is triggered by increased carbon dioxide in the blood, which makes your pH more acidic, and this leads to the sensation of breathlessness. This means that the more activity you do, the more your brain signals that you need to breathe more rapidly, and your heart beats faster to keep the exchange of oxygen and carbon dioxide happening. So the more efficiently you can deliver oxygen to your muscles, and the better you can tolerate increased carbon dioxide, the better your breathing system will be, the better your health in everyday life will be and the better your results in the gym will be.

Now, those *without* an effective oxygen delivery system, resulting from bad breathing habits, injury, illness or those with a low tolerance of carbon dioxide, are likely to experience a quick onset of the feeling of breathlessness during physical activity. You might even find it hard to walk to your local shop or get up a flight of stairs without feeling out of breath. So delaying the onset of breathlessness through building up carbon dioxide tolerance is important for us all, to live healthy, active lives. But there's another piece to the breathing puzzle.

If by breathing aerobically (using oxygen in the air) you cannot meet the energy demand required for a given activity or if your cells run out of oxygen, a new process starts. This is called *anaerobic respiration*. Anaerobic respiration does not need oxygen from the air. It works by releasing a relatively small amount of ATP (the energy source your cells use, which we discussed at the start of Chapter 1) during an internal fermentation process called *glycolysis*. This process creates a waste product, lactic acid, which makes your muscles ache and reduces your ability to perform, not just in your given sport but in your general breathing economy and health.

THAT WEE BIT EXTRA

Between 1908 and 2008, British cyclists won just one gold medal at the Olympic Games. As for the Tour de France, the world's biggest and most famous race, no British cyclist had won it in 110 years. But at the turn of the millennium something strange was happening. A maverick sports scientist and bike sales manager named Dave Brailsford was hired by British Cycling as a consultant in 1998 and became performance director in 2003. He pursued a bold new strategy based on the theory of marginal gains.

"The whole principle came from the idea that if you break down everything you could think of that goes into riding a bike, and then improve it by 1 percent, you will get a significant increase when you put them all together," he said.[59] So, while Dave's team focused on improving aerodynamics, they didn't stop there; they redesigned bike seats to make them more comfortable, they tested different types of massage gels to see which one led to the fastest muscle recovery, they determined the type of pillow and mattress that led to the best night's sleep for each rider, and they even painted the inside of the team truck white, which helped them spot little bits of dust that would normally slip by unnoticed but could degrade the performance of the finely tuned bikes. They considered anything and everything that could be made 1 percent more effective to give them that significant improvement. It's a great way to look at life and work in general, not just sport.

And it worked. At the 2008 Olympic Games in Beijing, British cyclists won 60 percent of all the gold medals on offer. At London in 2012, they set nine Olympic records and seven world records. That same year, a British cyclist—Bradley Wiggins—won the Tour de France. Next year, a different British cyclist—Chris Froome—won it. And then he won it again—in 2015, 2016 and 2017. That's five Tour de France wins in six years. Over a decade, from 2007 to 2017, British cyclists won a staggering 178 world championships, 66 Olympic or Paralympic golds, and five Tour de France victories.

So when I turned up in leafy Henley-on-Thames to work with some Olympic rowers, they, like the cyclists, were looking for that little bit extra—that 1 percent in their breathing—which, when combined with other tiny improvements, would make a big difference. As one of them put it, "A lot of our rhythm in a race is built on how we breathe. It's always been an aspect of our racing that's very dominant." Another agreed: "In an Olympic final there are 200 strokes, and a huge amount of work has gone into each and every one of those strokes, so if we can do one of those strokes even a point of a percentage better, all those things add up."[60]

Oxygen is the fuel for your muscles. In order for you to do *anything*—talk, walk or exercise—you need to get oxygen to your muscles. Breathing better would surely provide a marginal gain for the rowers. But what I soon found was that the potential for improvements was well above 1 percent.

I was struck by how some of these great athletes were breathing. One of the rowers was a chest-dominant breather. Another had lower-than-average oxygen saturation levels, which meant his breathing wasn't effectively fueling his muscles. What this shows is that even athletes at the top of their game can breathe in a way that limits their overall function and performance. And that's hugely exciting. Think about the kinds of superhuman performances we might be able to see if top athletes can really understand the power of breathing. It's possible that we could see not just marginal gains, but massive gains. And if professional athletes could be breathing better, what does that say for the rest of us?

The first stage for the rowers (and for you too) was to make sure their breathing mechanics are working efficiently. You have hopefully made some changes to your breathing by this point. But I want to think specifically now of how you breathe during physical exercise.

EXERCISE 36

BREATHING MECHANICS VISUALIZATION

I want you to imagine yourself exercising right now, and you could be doing anything at all: running, hiking, dancing, lifting weights—whatever it is you like to do.

- Picture yourself pushing yourself to maximum intensity.
- How are you breathing?
- Are you breathing in your chest or belly?
- Through your nose or mouth?
- How do you feel?

If it's hard to envisage it in your mind, pay close attention to how you're breathing when you next exercise. Remember, awareness is the first step for everything. Or, if you like, you could try running on the spot now or doing some star jumps for five minutes to see how you breathe.

NOSE, NOSE, NOSE!

As you've learned in previous chapters, the most effective way to bring air into your lungs is through your nose, using your diaphragm. In sports it's no different. And yet, if you're like most of the runners I see panting along the canal near my house, you'll be breathing through your mouth, especially as the intensity of activity increases. It happens naturally when we feel like we can't get sufficient air quickly enough for a given activity—our breathing speeds up, gets short and shallow, and we switch to mouth breathing. Mouth breathing provides less air resistance than nose breathing, the chest moving short and shallow so we can quench our thirst for air faster. Despite this apparent quick fix, research suggests that mouth breathing is not as economical a process for oxygen delivery.

This was explored in a study undertaken at Colorado State University.[61] Researchers looked at male and female recreational runners who had been practicing nose-only breathing while exercising for six months. Their VO_2 max, the maximum rate of oxygen consumption ("V" for volume, "O_2" for oxygen and "max" for maximum), did not change from nose breathing to mouth breathing. But their breaths per minute decreased during nasal breathing, meaning they could have just as much output by breathing less. Breathing less is more economical and less stressful for the system than a panting, hyperventilated breath.

As you now know, one of the main jobs of your nose is to support your respiratory system—preparing the air, filtering out particles, and adding moisture and heat to improve the entry of air to the lungs. Nitric oxide is also released during nasal breathing, and this increases blood flow and lowers blood pressure. If you're not breathing with your nose, it can cause complications.

During the routine screening of the UK Olympic team before the Athens Olympics in 2004, the recorded prevalence of exercise-induced asthma (a condition where your airways narrow, making it difficult to breathe) among the athletes was 21 percent—double the prevalence in the UK population.[62] This is quite incredible, and logically makes little sense. Why should exercise-induced asthma be so high in some of the nation's fittest people? It seems to be even more prevalent in winter sports: some studies show rates as high as between 35 and 50 percent in top figure skaters, ice hockey players and cross-country skiers.[63] So what's happening? Perhaps you can guess.

Scientists speculate that exercise-induced asthmatics have an increased tendency to switch to mouth breathing during their sport, which reduces moisture in the airway, causing dehydration and inflammation.[64] However, breathing through your nose provides a protective influence against this. Research has demonstrated that breathing only through the nose during exercise inhibits a restricted airway response during exercise.[65] Of course, there are some sports, swimming being the most obvious, in which it's impossible to practice nose-only breathing, but you can certainly retrain to use your nose in many sports.

This is something Patrick McKeown, CEO at Oxygen Advantage®, has been advocating for many years. As a disciple of Dr. Buteyko and The Buteyko Method, Patrick is one of the world's leading breathwork experts, who has spent over 20 years researching and teaching the science of optimal breathing, and in particular its application for sport. He has helped thousands improve their health and performance through his books and Oxygen Advantage training programs.

WHERE DOES FAT GO?

Ever wondered where weight goes when you lose it? Does it evaporate? Do you sweat it out? Does it go down the loo? Into the ether?!

There are widespread misconceptions about how humans lose weight. Fat can't just magically transform into heat or energy—that would go against the law of conservation of mass. Fat doesn't break up into smaller parts either. Nor does most of it get excreted, or turned into muscle, despite what some think.

So where does it go? You can guess what's coming. That's right, most of it is breathed out.

This is how it works:

The average human fat molecule's chemical makeup is $C_{55}H_{104}O_6$. That sounds like a robot from *Star Wars*, but what it means is that fat is made up of 55 carbon molecules, 104 hydrogen molecules and six oxygen molecules. When you move more and eat less, creating an energy deficit in the body, fat is "burned," which means it's broken down into these molecules: carbon, hydrogen and oxygen.

Now we already know that glucose and oxygen make ATP, the main molecule for storing and transferring energy in cells, as well as carbon dioxide and water. When there isn't enough glucose in the body to meet its demand for energy, your body goes to your fat cells to get it. This means that the majority of your fat, 84 percent of it, is breathed out as carbon dioxide, and the other 16 percent is sweated, urinated or excreted out as water.

I know what you're thinking. If I want to fit into those jeans, can I just breathe more? Sadly not. It's the energy deficit that makes the switch happen.

YOUR NOSE IS A SMARTWATCH

Switching to nasal breathing during exercise can be a challenge at first. It takes time for the body to adapt. We've already been practicing this with "Let the Nose Do Its Thing" (Exercise 6) in Chapter 3. Even our box breathing (Exercise 31) and Jedi breathing (Exercise 32) can weaken the feeling of air hunger. Knowing that air hunger is not the enemy, you can start practicing nose breathing during warm-ups and cooldowns for your workouts.

Then change your perspective on your activity. Let's say you're heading out for a run. Instead of thinking about your speed or time, just go as far as your nose breath will take you. *Only* breathe through your nose as you run. Whenever you're "out of breath" or feel the desire to mouth breathe, slow down to a walk until you catch your breath. Just keep nose breathing no matter what. If your nose is blocked, remember our nose-unblocking technique from Chapter 3 (Exercise 4). Over time you'll develop your ability to nose breathe while you run, and find it easier.

I like to use my nose breath almost like a smartwatch when I'm running, to signal when my body is trying to switch between aerobic and anaerobic respiration. This way you can start to work on developing your aerobic threshold, limiting lactic acid buildup, and preventing sore muscles. The restrictive nature of the nose breath tends to keep you in an aerobic state. When you feel that desire to mouth breathe, maybe when you reach a steep hill or you're going faster, you tend to switch to anaerobic. So play around with that threshold: find a pace where you're testing the desire to mouth breathe but not so much that you have to. And remember, if you need to catch your breath, slow down to a walk, but keep nose breathing only.

WE NEED MORE HEMOGLOBIN BUSES

Let's take a trip back to Chapter 3 and remind ourselves about the hemoglobin bus. Pumped by your heart, the hemoglobin bus carries oxygen from our lungs to our cells and won't allow oxygen to hop off at the cells that need it unless those cells already contain the appropriate amount of carbon dioxide. Each hemoglobin bus only has four seats. So if you've a limited amount of buses, then oxygen in the lungs won't make it on time to work, and the *anaerobic respiration* process will begin to make up the energy required. This creates lactic acid, and you start to feel sore and breathless.

So alongside nose breathing during exercise, if you can manufacture more hemoglobin buses and create additional lanes on the arterial highways so they can travel easily, you can then increase your capacity for aerobic respiration. The extra lanes on the highway are fairly simple, as harnessing nitric oxide through nose breathing will help open blood vessels and add space to the highways. You could harness more by our humming technique too, which you practiced as part of the Recognize–Breathe–Reframe technique (Exercise 21). But to manufacture more buses requires a bit more intervention. You need to create more red blood cells, as red blood cells contain hemoglobin.

(DON'T) BE LIKE LANCE

There's a naturally occurring hormone in the body that stimulates the bone marrow to produce more red blood cells called erythropoietin, pronounced "ah-rith-ro-poy-tin" (EPO). More red blood cells mean more hemoglobin buses are available to transport oxygen to work on powering your muscles. Medically, EPO is given to patients with chronic diseases that suppress the bone marrow, and it helps them have more energy and increase daily function. However, EPO was thrust in the limelight for a different reason when disgraced cyclist Lance Armstrong admitted to Oprah Winfrey that he had illegally injected EPO, as well as having taken a cocktail of other performance-enhancing drugs, during *all seven* of his Tour de France victories as a way to boost his aerobic capacity and cheat his way to the podium over and over again. If only Lance hadn't taken this shortcut and instead worked to naturally (and legally) achieve the performance-enhancing power of EPO by working with his breath.

Your body naturally increases EPO to help support delivery of oxygen when you're in a low-oxygen environment. This is why you may have heard of athletes training at altitude. Even though intense training at altitude is difficult, if an athlete also sleeps at altitude, the increased EPO effect dramatically increases performance when they compete at sea level. Thus came the development of hypoxic tents (*hypo* means "low" and *oxic* means "oxygen"), in which an athlete could sleep or lounge around for hours on end, stimulating the production of EPO. As EPO increases in the body, so do red blood cell counts and an athlete's oxygen-carrying capacity. This gives them an edge when stepping out and performing at sea level. It's a perfectly legal strategy and accepted by the World Anti-Doping Agency (WADA). Unless money is no object, however, it's unlikely that you're going to take this approach to give you the extra energy to play with your kids or improve your PB on your local park run. The good news is that a similar outcome can be achieved with a slightly different approach.

LIVE LOW, TRAIN HIGH

Another way to increase EPO is to live at sea level but train at altitude, which, as I've just mentioned, is physically difficult. (A further sticking point: you might not have access to altitude.) However, you can simulate altitude through hypoxic breathing techniques and hypoventilation—simply put, breath holding and retention. Before we explore this, it's important to clarify that practicing breath holds consciously is different from the unconscious breath-holding patterns that we explored in previous chapters. This is about having a *deliberate* breath-holding practice to evoke positive change in the body; it's not about not making a habit of it in your daily life.

One of the pioneers in this field was a man once dubbed the "Greatest Runner of All Time" and the "Czech Locomotive," Emil Zátopek. He only started running because managers at his shoe factory forced him into a race. And that wasn't the only unconventional thing about him. As Larry Snyder, track coach for Ohio State University at the time, said, "He does everything wrong, but wins."[66]

Just like my judo coach Billy made us do, Zátopek would often hold his breath while training, something considered strange at the time but now supported by modern science. He understood that if he could train hard while taking in less oxygen (hypoventilation), his body would have to get more efficient at using it. And that's exactly what happened.

Zátopek broke 18 world records and went undefeated for six years at the 10,000 meters.

There are two ways to carry out voluntary hypoventilation: either at high pulmonary volume—you breathe in and hold your breath with full lungs—or at low pulmonary volume—you breathe out and hold your breath with empty lungs. Hypoventilation at high pulmonary volume, used for decades by swimmers or runners like Zátopek, provokes an increase in carbon dioxide concentrations. However, further studies conducted by the Université Paris 13 have shown that hypoventilation at low pulmonary volume can provoke both an increase in carbon dioxide concentrations (hypercapnic effect) and a drop in oxygen (hypoxic effect) in blood and muscle, simulating being at altitude to promote an increase in EPO.[67] When an exhale-hold technique is correctly applied, it's possible to obtain a decrease in blood oxygenation similar to what would occur at altitudes above 2,000 meters.

INCREASE YOUR PERFORMANCE ON THE FIELD

In many sports such as rugby or football, the ability to sprint repeatedly and recover is an important fitness requirement. Xavier Woorons of the University of Lille ran a four-week study on 21 highly trained rugby players to determine the effects of repeated sprint training in hypoxia induced by voluntary hypoventilation at low lung volume.[68]

The players performed repeated 40-meter sprints every 30 seconds until they dropped below 85 percent of their maximum speed for a sprint. The subjects were then matched into pairs for performance level and randomly assigned into groups: one that performed the repeated sprint training as normal, and one that performed exhale-hold sprints (or intermittent hypoxic training). Woorons and his co-researchers concluded that the number of sprints completed by the hypoxic training group was significantly increased by the end of the trial. After four weeks this group improved from 9.1 sprints to 14.9 sprints, while those who trained normally showed only a slight improvement, from 9.8 to 10.4 sprints. Maximal velocity did not change, just the ability to keep going at that pace for longer, showing it to be an effective strategy to improve repeat sprint running in team-sport players.

BEYOND PERFORMANCE

Understanding how the body responds in low-oxygen environments through practicing deliberate breath holding has benefits beyond physical performance.

Stem cells, for instance—which play a crucial role in delaying the aging process—have been shown to survive longer and renew themselves in a low-oxygen environment.[69] Conscious breath-holding practice has also been shown to induce p53, the so-called "guardian of the genome,"[70] which is considered the most significant protein for cancer suppression.[71] Loss of function of p53 underpins tumor progression in most epithelial cancers, which account for 80 to 90 percent of all cancers, according to the National Cancer Institute.[72] (Epithelial tissue is found throughout the body; in the skin, as well as in the covering and lining of organs and internal passageways, such as the gastrointestinal tract.) Deliberate breath-holding practice, not unconscious breath-holding patterns, can also help further increase the production of nitric oxide, which, alongside the benefits already mentioned, can help the body's defense mechanisms against oxidative damage. This clearly illustrates the power of breathwork to improve your physical health right down to the cellular level.

Once again, research is starting to support what traditions have claimed for centuries. Researchers have also found that deliberate breath holding for short, intermittent periods may increase life span by preserving the health of stem cells.[73] It can also cause the body to release stem cells to repair major damage, and prevent and relieve disease such as Parkinson's.[74] Breath-holding practice can also preserve brain function[75] (this is theoretical in humans, though; studies have only been done on salamanders) and can help increase resistance to bacterial infections.[76]

The relationship between breathing and immunity was poorly understood until one man convinced of breathing's power set out to rewrite science books—and achieved his aim.

Deliberate breath holding for short, intermittent periods may increase life span.

WIM "ICEMAN" HOF

"Breathe in, breathe out. Just go with the flow. IN, OUT, IN, OUT. Into the belly, into the chest and let it go. Like a wave. That's it, fully in…let it go…just keep on going, keep breathing. Fully in…let it go… Give it all you've got. Breathe in your ass if you have to. Last one…fully in…and let go. Breath hold from now on. Let the body do what the body is capable of doing… Relax to the deepest, whatever it takes…"

This was the first time I'd met Wim Hof, teaching a packed-out room in London. Wim is a refreshing addition to the world of breathwork. He shook things up and has shifted the perception of breathwork from New Age fad to mainstream practice. Here in this packed room were mainly rugged men: rugby players, PTs, fitness nuts and A-types in gym clothing, people that liked to push their body and mind to the limit, always looking for something to give them an edge.

This is what Dutch daredevil Wim does. He pushes the boundaries of what's humanly possible. In some ways, he's like a modern Swami Rama, yet more accessible. He claims that whatever feats he's capable of, you can do too. He teaches people how.

Wim Hof earned the nickname "Iceman" after performing a number of stunts and setting several world records. He swam under a frozen lake on live TV, then jumped right back in to rescue an onlooker who had fallen into the freezing water. In 2007 he climbed to 24,500 feet on Mount Everest with no thermals, supplementary oxygen or goggles, and wearing nothing but a pair of shorts and open-toed sandals. He also climbed Mount Kilimanjaro in similar attire, and ran a half-marathon above the Arctic Circle barefoot. He nearly died when his retinas froze over after swimming 50 meters under the polar ice and he had to be rescued. When people thought it was just the cold he could endure, he completed a marathon in the Namib Desert in Southern Africa—without drinking any water during the run.

"You use stress to find the goal," he claims. His teacher? Cold and hard nature. His secret? "Breathe, motherfucker!"

Wim uses the Tummo breathing technique that Tibetan monks practice to heat themselves up, alongside breath holds and cold exposure. He has developed what he calls the "Wim Hof Method," and it has gained a large following around the world. This method combines breathwork and cold exposure to force and shock the body. You hyperventilate for 30 deep breaths, hold your breath for as long as you can and, after some rounds of this, jump into iced water while controlling your breathing. His method is about using the natural sympathetic stress response of the body in a controlled way to push the boundaries of what's humanly possible.

One of the most profound uses of this technique to give yourself so-called "superhuman powers" has to do with the autonomic nervous system and the immune response. Wim claimed he could deliberately influence his autonomic nervous system and access his immunity. So under the watchful eye of Dutch researchers, he was injected with an endotoxin. He then practiced his breathing technique and, as he claimed, dampened his immune response so that he did not get sick or experience any adverse symptoms.

Amazed by this, researchers considered whether Wim's genes were different from those of the average person, giving him the ability to withstand endotoxins in his blood. But Wim claimed otherwise, saying that it was possible for anyone to do this. So a further study was conducted with 30 healthy males, 18 of whom trained in his method, the other 12 being a control group who had no training. All participants were given the endotoxin—all the control group exhibited the typical symptoms and reactions of the immune response (fever, nausea, headaches and shivering), while those who trained with Wim showed no acute symptoms, had reduced inflammation and recovered quickly.[77]

BREATHING FOR IMMUNITY—THE WIM HOF STUDY

When toxins invade your body, they trigger an innate immune response, which gears up to fight off the infection. As the immune response kicks in, it releases small proteins called cytokines and you experience physiological changes: fever, nausea, shivering, headaches, etc.

The combination of deliberate intermittent hyperventilation breathing and strong breath holding alters the body's chemistry, triggering a stress response and flooding the body with adrenaline as a result. This has been shown to boost interleukin 10—a key messaging protein that acts as an anti-inflammatory cytokine, inhibiting the release of the other cytokines that lead to inflammation. Even though the inflammatory response is suppressed, the increased adrenaline triggers an increase in white blood cells in the blood available to fight off the toxin.

So the immune system works in the background, despite the dampened inflammation response. This shows that you can deliberately hack into your autonomic nervous system and use an acute stress response to fight off toxins or infections without experiencing acute symptoms. The result is you recover faster. More research is needed, but this could be a promising alternative to traditional treatment for those with inflammatory or autoimmune disease.

In some ways, what Wim Hof tries to experience are the extremes that we'd endure in nature. These would trigger acute stress responses, which enable our bodies to do things we never thought humanly possible.

Modern human life is fairly luxurious. If we're cold, we turn the heating on; too hot, on goes the air con. As a species, with the advancement of technology and the development of our brains, we've become disconnected from the natural strength that got us where we are today. We often live our lives half asleep—or half awake. Through more extreme breathing practices and by exposing ourselves in a controlled way to physically stressful situations while keeping a calm state of mind, we can force our body to reset and adapt, altering both our body and mind.

The story of Wim Hof is of someone who made the impossible possible through breathing.

TAKE YOUR RUNNING TO THE NEXT LEVEL

Helen was effortlessly keeping her pace and gliding toward the finish line. Sweat was dripping down her face, which was red with the exertion, but she looked comfortable. Her running was smooth, her breathing calm. And she started to grin. She'd done it—completed the Derby Half-Marathon and raised £5,000 for the Alzheimer's Society. *And* she'd beaten her personal best by 15 whole minutes.

How did she do it? Helen, a primary school teacher and mum of two, had been following a simple breathwork for runners plan I'd shared with her to help improve her health and fitness. Central to this was something called *rhythmic breathing.*

EXERCISE 37

RHYTHMIC BREATHING FOR RUNNING

It goes without saying that rhythmic breathing involves you breathing using your diaphragm and practicing nose-only running, which we covered earlier in the chapter.

Next you've got to find a rhythm that matches your breathing to the cadence of your footsteps. This can be a helpful way to find your flow, keep centered and stay focused while you run. But make sure it's an odd/even pattern to reduce impact stress; landing on the same foot at the beginning of every exhalation causes one side of your body to absorb the greatest stress and impact of running.

This is because when your foot hits the ground, the force of impact equals two to three times your body weight. This impact stress is at its greatest when your foot hits the ground at the beginning of an exhalation because as you breathe out your diaphragm relaxes, creating less stability in your core. Less stability at the time of greatest impact creates a perfect storm for injury, especially if it's repeatedly happening on the same side. Rhythmic breathing, on the other hand, with an odd/even pattern, alternates the impact stress across both sides of the body when running to reduce your chance of injury.

There are a couple of rhythmic breathing patterns I recommend, both of which have a longer inhalation than exhalation. Why? Because your diaphragm and other breathing muscles contract during inhalation, which brings stability to your core. These same muscles relax when you breathe out, decreasing stability. With the goal of injury prevention in mind, it's best to hit the ground more often when your body is at its most stable: during inhalation.

Start with a 3-2 pattern of rhythmic breathing, which will apply to most of your easy-to-moderate-effort running. Inhale for three steps and exhale for two.

To get the hang of this, practice first on the floor:

- Lie on your back with your knees bent and feet flat on the floor.
- Place a hand on your belly and make sure that your diaphragm is engaged, so you can feel it rise and fall.
- Breathe in through your nose for three counts, adding foot taps to mimic the steps.
- Breathe out for two counts, adding foot taps to mimic the steps.
- In (*left*), 2 (*right*), 3 (*left*), out (*right*), 2 (*left*), in (*left*), 2 (*right*), 3 (*left*), out (*right*), 2 (*left*) and so forth.

If you need to pick up the pace, or you have a hill to get up, then switch to a 2-1 pattern: breathe in for two steps and breathe out for one. This is also helpful for staying injury-free when interval training and sprinting.

EXERCISE 38
BREATH-HOLDING PRACTICE FOR SPORT

Rectangle breathing is a good place to start building your breath-hold length. A cousin of box breathing, it will help you to engage your diaphragm while introducing you to breath holds at both high and low pulmonary volume. The in- and out-breath remain constant, but your holds increase over time. This exercise will also promote a focus response, which is another bonus.

- Sit in a comfortable position or lie flat on the floor.
- Relax your shoulders.
- Put both hands on your lower abdomen.
- Breathe in through your nose and into your hands for a count of four, feeling your belly rise.
- Hold your breath for a count of five. Keep calm and relaxed. (Try not to tense your muscles when you're holding your breath.)
- Breathe out through your nose for a count of four, feeling your belly fall.
- Hold your breath for a count of five. Keep calm and relaxed.
- Repeat for five minutes.
- Keep the in-breath and out-breath at four seconds, but each week try to increase the hold size by one count. Don't rush the exercise. It will take time for your body to adapt.

Example:

- Week 1: Practice IN four, HOLD five, OUT four, HOLD five. Repeat five minutes a day
- Week 2: Practice IN four, HOLD six, OUT four, HOLD six. Repeat five minutes a day.
- Week 3: Practice IN four, HOLD seven, OUT four, HOLD seven. Repeat five minutes a day.
- Week 4: Practice IN four, HOLD eight, OUT four, HOLD eight. Repeat five minutes a day.

You may feel a desire to breathe more, but try to resist it. This is just carbon dioxide building up in your body. What we're aiming to do is build your tolerance, so try to keep calm and work through it. If you want to challenge yourself further, you can practice this when walking.

DIVE DEEPER FOR THAT 1 PERCENT MORE

There's a correlation between the positive effects altitude has on breathing economy and the physiology of the deep-sea-diving populations scattered round the world. This last little deep dive could be the final 1 percent you need. The Ama Pearl divers in Japan, and the Bajau subsistence fishers in the Philippines and Malaysia, do repeated dives deep into the ocean, spending as much as 60 percent of their time underwater. Over the course of a nine-hour day they might spend as much as five hours underwater not breathing.[78] These diving populations share distinctive characteristics with those who live at altitude, such as the Sherpas in Nepal, known for their efficient breathing. They have big, strong lungs, great lung capacity, lots of red blood cells (because of the production of EPO in low-oxygen environments) and larger-than-average spleens. The spleen's role in oxygen delivery could be the final key to improved oxygen delivery, especially during sporting performance.

Your spleen stores about 25 to 30 percent of your concentrated red blood cells. Seals (real seals this time, not US Navy SEALs) are some of the animal kingdom's most impressive divers, storing about half their red blood cells in their spleens. This means they don't waste energy pumping all that extra blood around their body when it's not needed. When you practice a strong breath hold (or even just do a hard workout), your spleen contracts, much like a seal's, to release additional oxygen-rich blood into your circulation to sustain energy. With breath-hold training, this contraction can be stimulated to nudge better performance by providing enhanced availability of oxygenated blood.

In the case of Sherpas and the Bajau free divers, their genes are responsible for bigger lungs and bigger spleens, thanks to the time their ancestors have spent either high in the mountains or underwater. But it's not necessarily the spleen's size that matters—its squeeze response that enables red blood cells to move into circulation is vitally important, and this is something that can be trained to help with your performance. It requires a slightly more advanced breath-holding practice that mimics deep-sea diving but in the comfort of your own home. These are dry-land exercises, and should not be practiced in water, in a bath or if you're pregnant.

There are two types of breath-holding practice I like to use that have slightly different outcomes. The first practice helps improve your tolerance to high levels of carbon dioxide to help with EPO production and resetting your breathlessness clock. The second practice helps your ability to be in low O_2, and increases spleen contraction and stem cell longevity. Deep-sea divers call these the CO_2 and O_2 tables, as they mimic diving but can be done at home. Divers tend to practice their breath holds on full lungs, but I like to practice the holds on the tables on empty lungs as you get similar results in less time.

They both consist of holding your breath eight times. With a CO_2 table, the length of time you hold your breath stays the same, but the recovery time between each round decreases each round. With O_2 tables, your recovery time stays the same but the length of your breath hold increases each round. You can calculate your own as shown in the exercise, or there are apps that do this for you.

EXERCISE 39

PERFORMANCE DEEP DIVE

STEP 1: CALCULATE YOUR MAXIMUM BREATH HOLD

Calculate your maximum breath-hold time on empty lungs (low pulmonary volume). You'll need a stopwatch for this.

- Find a comfortable spot, seated or lying down.
- Breathe calmly for two minutes to relax the body and mind. Be mindful not to over-breathe.
- After two minutes is up, take a normal breath in, a normal breath out, and start the clock to time your maximum breath-hold time. Keep calm, and your body will override your mind and make you breathe when you have to.
- Mark your score, as you'll need this for the O_2 and CO_2 table practices.

STEP 2: DECIDE WHICH PRACTICE YOU WANT TO WORK ON—PRACTICE 1 (HIGH CO_2) OR PRACTICE 2 (LOW O_2)

These tables are not here to push your limits; they're to condition you, which takes patience and time. This is about taking baby steps, and making slow and steady progress in a safe environment.

Things to note:

- Practice sitting or lying down.

- Do not practice in water or in a bath.
- Do not do more than one practice a day.
- Avoid going for maximum breath-hold attempts on the same day.
- Make sure you do not over-breathe or heavy breathe in your recovery rounds, as that will reduce carbon dioxide and defeat the object of the practice.

PRACTICE 1: HIGH CARBON DIOXIDE PRACTICE

People who have an early onset of breathlessness in sports need to concentrate on CO_2 tables. The idea of this is to let the level of carbon dioxide in your blood and tissues creep up slowly throughout the exercise. There are eight rounds of breathing and eight breath holds. The length of time you hold your breath stays the same (half your maximum), but the recovery time between each hold diminishes. Start with 20 breaths and work down to six breaths. Because there's less time in between breath holds to release carbon dioxide from the body, the amount of carbon dioxide in your body gradually increases with each repetition. This slow creep develops your tolerance to carbon dioxide and increases EPO. Practice every other day for two weeks, before trying the O_2 table. Do not do it on the same day as your max breath hold.

Breathing in recovery rounds should be natural and relaxed. Do not over-breathe, as that defeats the point of the training.

Example: if you calculate your maximum breath hold as one minute, your CO_2 table breath hold is set to 50 percent of that, so 30 seconds.

- Round 1: Breathe normally for 20 breaths, exhale and hold 30 seconds.
- Round 2: Breathe normally for 18 breaths, exhale and hold 30 seconds.

- Round 3: Breathe normally for 16 breaths, exhale and hold 30 seconds.
- Round 4: Breathe normally for 14 breaths, exhale and hold 30 seconds.
- Round 5: Breathe normally for 12 breaths, exhale and hold 30 seconds.
- Round 6: Breathe normally for 10 breaths, exhale and hold 30 seconds.
- Round 7: Breathe normally for eight breaths, exhale and hold 30 seconds.
- Round 8: Breathe normally for six breaths, exhale and hold 30 seconds.

PRACTICE 2: LOW O_2 PRACTICE

This second practice is designed to increase your maximum breath hold, by increasing the amount of time that you hold your breath on each round. The recovery time stays the same. This trains the spleen squeeze, helps stem cell preservation and release, and induces p53. You start with relaxed diaphragmatic, nose breathing for 20 breaths. Increase your breath hold each round until the final Round 8 is 80 percent of your max. Breathing in your recovery rounds should again be natural and relaxed. Do not over-breathe.

Example: if you calculate your maximum breath hold as one minute, your O_2 table maximum breath hold is set to 80 percent of that—48 seconds—then work backward.

- Round 1: Breathe normally for 20 breaths, exhale and hold 13 seconds.
- Round 2: Breathe normally for 20 breaths, exhale and hold 18 seconds.
- Round 3: Breathe normally for 20 breaths, exhale and hold 23 seconds.
- Round 4: Breathe normally for 20 breaths, exhale and hold 28 seconds.
- Round 5: Breathe normally for 20 breaths, exhale and hold 33 seconds.

- Round 6: Breathe normally for 20 breaths, exhale and hold 38 seconds.
- Round 7: Breathe normally for 20 breaths, exhale and hold 43 seconds.
- Round 8: Breathe normally for 20 breaths, exhale and hold 48 seconds (80 percent of max).

Remember not to push your limits too far by either setting your CO_2 table hold at over half of your maximum hold or having your last O_2 table hold at more than 80 percent of your maximum hold. Remember that practicing these is like exercising your lungs, so view them as exercise; this means don't push your lungs physically with other physical exercises on the days you practice the tables, as you need time to rest and recover in between.

Every couple of weeks, check your max breath hold and adjust the tables. You may also like to see how much your carbon dioxide tolerance test number improves (Exercise 12).

In this chapter we've discussed how you can use your breathing to give yourself an edge, whether you're a pro athlete or someone trying to beat their PB. We've explored how the nose is crucial to this, and how you can use it like a built-in smartwatch to work within your aerobic threshold. You also now know how to push your deliberate breath holds to improve your given sport that little bit more. And you've learned about how breath-holding practice can provide benefits beyond the sports field with Wim Hof, who's demonstrated that there's a powerful link between breathing and immunity.

The story of Wim Hof is of someone who made the impossible possible through breathing. And what I've learned on my journey is that breathing can do this in a number of ways.

10

ACCESS TRANSCENDENT STATES

CELESTIAL CANOE TOURS LTD

"I was in a canoe, in what I can only describe as an enchanted tunnel. It was like one of those ones you go through on a log ride at a theme park, before it opens out into the world."

This was Ryan, a British film director and cinematographer. He was describing the experience he'd had during his breathwork session with me.

> I could see a light up ahead, but there was no need to paddle toward it. There seemed to be a current moving me gently along. I started to hear voices rising—"Keep going, that's it, you're almost there." As I passed through the opening out into a vast space—whoosh!—I found myself in com-

plete silence. I'd arrived in what felt like a giant underground lake illuminated by bright starlight. There was an overwhelming stillness to the space and yet something magical about it. It seemed to sparkle and shine. I felt a deep peace I'd never experienced before in my life. Yet this feeling also felt strangely familiar—so familiar that it made me emotional.

The water felt like silk against the paddle. We continued on easily, patiently, without expectation. And I say "we" because out of nowhere another canoe appeared just over my left shoulder and behind me, although when it appeared I realized it had been with me all along. On it was a man. I couldn't see his face, but he wore a black leather jacket and had long curly brown hair. He cruised past. It was weird. I felt like I'd known him all my life, perhaps longer.

"Excuse me, where are we?" I asked.

"Welcome, man," he said. "You've arrived. We've been waiting for you. This is the place where the universe comes together, where you can make sense of the world. Congratulations."

I know, I know. It sounds mad. But being transported to this other world, to this vast cave filled with starlight, where the water glittered, was one of the most powerful, calming and fulfilling experiences I've ever had. It gave me a better understanding of who I am, and a connection to a deep part of myself I'd hidden away. Not only that—it gave me a sense of connection to something bigger at play in my waking life and a clarity I'd never had before.

The feeling I experienced in this realm became the defining reference point for me. It enabled me to be able to access deep meditative states on a daily basis. I often need references in order for me to execute something; and I'm a perfectionist who likes to push myself. I'd never been able to meditate before. I always found it hard to just be, and simply gave up; but this breathwork experience, as strange as it was, helped me to understand the nature of "being" and what peace truly feels like. I was able to start practicing to connect to and experience more of this feeling in my waking life and creative work.

Sound strange? The funny thing is, these kinds of stories are really not uncommon. At the end of my breathwork sessions people often describe having had bizarre visions, leaving their bodies and traveling to another realm, experiencing an awareness of something much bigger than they do in their normal waking life or accessing a deep serenity that they've never felt before.

I can think of many occasions when someone in one of my groups has said they've felt the emotion that someone else in the group was going through as if that feeling were theirs. I've had countless accounts of people connecting with loved ones that have passed away. And time and again at the end of a group session, someone says something like: "It was so comforting when you held my head and someone else came and held my feet at the end of the session"—when neither I nor any of my facilitator team have done so. You may have even experienced something inexplicable in your daily infinity practice…

CONNECTING WITH SHAKESPEARE

Breathe in, and a part of the world becomes a part of you. Breathe out, and a part of you becomes the world. Breathing transcends your body and mind, connecting you to everything living. Air passes through your lungs and heart when you breathe in, and flows on for someone else to breathe when you breathe out. And of course, it's not just human beings that breathe the air around you—your cat breathes it, your neighbor's dog breathes it, the tree outside breathes it, the plant in your room breathes it and the oceans too. So when you breathe, you quite intimately connect to everything, and in a way your heart touches the hearts of a million others with each breath. It's like there's an invisible web of breath connecting us all.

When I say that breathing connects us intimately with everything else, I'm not exaggerating. In fact, as you breathe in and breathe out right now, you've just breathed the same air molecules inhaled (and exhaled) by a vast range of people, from Oprah Winfrey to William Shakespeare. That's right—in the course of a single day, you breathe the same oxygen molecules breathed by every living being that has ever inhabited the planet. You have a bond with every animal, every plant and every human that lives and has ever lived on earth every time you take a breath. So take another deep breath in. This is a powerful reminder of our interconnectedness with everything around us. The very mechanism that keeps us alive—breathing—also binds us to our fellow creatures and our planet.

YOU BREATHE THE SAME AIR AS EVERYONE ELSE, EVER

Breathe in.

You just breathed the same air as everyone on earth, ever.

For my fellow math geeks, here's how:

You breathe in about 25 sextillion molecules every time you inhale. (That, by the way, is 25 with 21 zeros after it, so quite a lot.) There are 7.75 billion people alive today. If each one of us had 7 billion descendants—so, that's 7 billion × 7 billion—you'd still be short of that number by 500 times.

So how long do these molecules persist in and get distributed around our atmosphere? The math for this is a lot more complex, but stick with me. The air is (roughly) a mixture of four molecules of nitrogen to every one of oxygen. So the mass of one mole of air—a mole is the base unit of the amount of substance—is about 28.9 grams. The total mass of our planet's atmosphere is about 5 × 1,021 grams. One mole of any substance contains about 6 × 1,023 molecules. Therefore, there are about 1.04 × 1,044 in the earth's atmosphere. Still following?

A mole of any gas at body temperature and atmospheric pressure has a volume of about 25.4 liters. The volume of air breathed in or out in the average human breath is about one liter. William Shakespeare, Napoleon, Oprah Winfrey—

whoever you like—will breathe out about 2.4 × 1,022 molecules per breath.

Let's take a random figure from the past. Over 45 years, that person, breathing at the average rate of about 25 breaths per minute, will breathe out about 2.1 × 1031 molecules. Therefore, every one molecule in every 5 × 1012 molecules in the atmosphere was breathed out by this person. But we breathe in about 2.4 × 1022 molecules each breath, so, even though we're making crude assumptions, there's an extremely high possibility that we breathe in about 4.3 × 109 molecules that this person breathed.

**When you breathe,
you quite intimately
connect to everything.**

A HARD WORD TO DEFINE

Powerful feelings of connection to something otherworldly, as in Ryan's breathwork experience, are often described as "spiritual." Breathing has traditionally been connected with spirituality; in fact, the English word *spirit* comes from the Latin *spiritus*, meaning "breath." The Book of Genesis, the first book of the Hebrew Bible and the Christian Old Testament, explains how God breathed life into the nostrils of Adam, the first man. Breath is the force that sustains life in yoga, which Hindus believe brings them close to Brahman, the Ultimate Reality or Supreme God.

However, "spirituality" is a notoriously hard word to define, and it's loaded with all sorts of assumptions and beliefs. Some people are comfortable with this, others less so. One of the simplest and most inclusive ways to define "spiritual" is as something dealing with the nonmaterial world—things that aren't physical, that you can't touch. It also involves the recognition of a feeling or sense that there's something more to being human than our sensory experience, that there's a world beyond our dimension and that all of nature is connected to it in a divine way.

We know from quantum physics that a whole world exists at the subatomic level that we can't see or touch. Science is increasingly showing us that the world is more complex and mysterious than we imagine. For most people, the idea of spiritual experience is far less confusing and helps make sense of the world. In fact, people have been having spiritual experiences since the dawn of time, sometimes spontaneously, sometimes induced through ceremonies, rituals or even by simply being in nature.

With spiritual experiences it's very common to hear people talk about having a strong feeling that they're connected to something greater, that they have a sense of everything, that for a moment they seem somewhere beyond space and time. And with this can come a lot of positive emotions: love, joy, bliss, feelings of kindness and oneness, a deep sense of meaning. It's through these spiritual experiences that people change and transform, have profound intuitions, become more creative, understand themselves better, develop more trust in life, heal, solve problems, lessen their depression and harness flow states. Experiences like these can help people accept a big upheaval or a painful loss, like losing a loved one, and move on. This was certainly the case for me.

Really, you can call experiences like these whatever you want. But most people—even those who find the term "spiritual" uncomfortable or confusing—feel like they need this type of deep connection from time to time, whether for gaining a sense of purpose, meaning or acceptance, or just to reach an understanding about an aspect of the world. We all need to get back in touch with why we've chosen the path we've chosen. Sometimes we need a way to stand back and make sure the path we've chosen is the right one for us. I can think of many times when a client of mine has had a powerful experience through breathwork and realized that some area of their life—maybe a toxic friendship or a job they no longer love—isn't working for them.

Billions globally who have embraced or been born into traditional religions find their sense of spirituality is nurtured that way. And "spiritual, but not religious" has become a popular way for a person to self-identify as someone who has a strong sense of this connection beyond the physical, but doesn't recognize organized religion as the only path to growing spiritually. Some adopt more of an *à la carte* approach to religion, taking practices and teachings from other faiths, beliefs and traditions. And plenty of people lead deeply spiritual lives or have intensely spiritual experiences without meaning to, and without thinking of them as "spiritual." The point is that we all have something in us that seeks a connection to something beyond our material world, beyond our individual human experience. We all have an innate desire to connect or reconnect to it throughout our lives, and breathing practice may be a way to accelerate that connection.

LOOKING INWARD TO LOOK OUTWARD

In the case of some practices like breathwork and meditation, by looking inward you end up looking outward. You're able to find a source of healing, connect to your deeper intuitions, and find purpose and meaning in your life. Accessing these states, which some would call transcendent states, can be accelerated by either under-stimulation or overstimulation techniques. Slow breathing, meditation, prayer, hypnosis, fasting and other activities can achieve these effects. For thousands of years monks, nuns and yogis, among others, have practiced removing as many sources of stimulation as they can, while lowering the intensity of those that resist removal. In more recent times people have used sensory-deprivation tanks, or "silent retreats," to access these states. Under-stimulation encourages introspection and an understanding of how the part of the mind that gives us our sense of identity and strives after things—the ego—is the cause of much of our unhappiness and separates us from others. After a period of under-stimulation the ego drops away, and we experience both bliss and wider awareness.

On the flip side, radically amping up stimulation can also provoke a connection to a powerful and transcendent state by diminishing or eliminating the ego and its defenses. Activities that many cultures have used to do this include accelerated breathing, rhythmic stimulation (such as music, drums, rattling, dancing and singing, and chanting), sleep deprivation, heat and cold exposure, and plant medicines.

These practices or medicines create temporary changes in neural activity. These could be a change in your neurochemistry, how your neurotransmitters function, a change in the receptors in the brain or a change in brain waves, as explored in Chapter 5. These changes can make your brain less constrained, strengthen and create new connections within it and lead to an increase or decrease in certain chemicals, which allows for a change in consciousness and perspective. Spiritual experiences tend therefore to affect our emotions, vision and sense of bodily integrity.

SPIRITUAL BLISS

Yogis believe when the physical breath stops in meditation or breathwork, something they call *kumbhaka,* they can access a state of bliss, increasing vitality and positive energy in the body and mind. They consider this the final stage of spiritual union, or *samadhi.*

Kumbhaka is neither deliberately holding your breath nor an unconscious breath-holding pattern. It's an unprompted stopping of breath that occurs in the *samadhi* state, attained through meditation or breathwork. I've seen this in sessions: clients will stop breathing and report back that they experienced complete peace and bliss. It's claimed that in this state of *kumbhaka,* yogis experience enlightened insights, can cure diseases, boost health and increase longevity.

NON-ORDINARY: GROF'S RESEARCH

As you'll have noticed by now, the kind of experiences I'm talking about vary in intensity. At one end you might feel a stronger-than-usual sense of connection with the people around you, accompanied by positive emotion. At the other you might have a healing, life-changing experience that sets you on a brand-new path in life.

One man with a deep interest in what exactly is happening when we have experiences like these is the Czech-born psychiatrist Stanislav Grof. He has done extensive research into how you can safely induce a deep experience for the purposes of healing and gaining insight through the use of various methods. He called these experiences "non-ordinary" states of consciousness, and over 60 years he has published 160 articles on his findings.[79]

Unlike other, more conventional forms of therapy, Grof's focus was on accessing the unconscious mind experientially, rather than intellectually, verbally or analytically. His clinical research, first at the Psychiatric Research Institute in Prague from 1960 to 1967 and then at Johns Hopkins University, was extremely promising. Grof observed and reported remarkable therapeutic benefits of using the powerful hallucinogenic drug LSD for deep healing with psychiatric patients, cancer patients and drug addicts, as well as showing its wide-ranging creative benefits for artists and scientists curious about exploring the depths of their minds. But LSD was made illegal in the US in 1966 after a sharp rise in its recreational use and Grof's clinical research came to a halt.

Grof then turned his attention to cultures that had found ways to elicit similar experiences without the use of psychedelic substances. He studied drumming, meditation, fasting and chanting, and soon came across an especially direct and accessible way to access a non-ordinary state of consciousness: breathing. He found that rapid, equal breathing could bring subconscious experiences to the surface, initiating cathartic emotional releases and shifts in awareness. This style of breathing, which he called *holotropic breathwork*, could induce states that promoted inner healing and enabled the deep exploration of the human psyche. It was another form of conscious energy breathwork that appeared at around the same time as Leonard Orr's rebirthing technique.

To an outside observer, holotropic breathwork sessions can look pretty strange. There's mouth breathing, a lot of shouting, odd postures and weird movements, all designed to help the participants express and release trauma that they find hard to address through other means. The bizarreness of the practice is really an indication of how stubborn certain kinds of trauma can be to release, and I've seen firsthand quite how effective it can be. As long as the practice is done in the company of a qualified practitioner, it can provoke powerful transformations. Let me explain how.

BREATHE AWAY THE EGO

Prolonged rapid breathing and mouth breathing creates a state of hyperventilation. This hopefully raises some red flags with you. We've explored in earlier chapters how daily mouth breathing and hyperventilation is extremely dysfunctional, as it keeps the body out of balance, provoking a stress response. It's one of the things I've been wanting you to avoid and fix at all costs. But when used sparingly as a deliberate intervention for healing purposes, and if it's done in the presence of a trained practitioner, something pretty special happens. We enter a state of being that helps us let go of some of the control patterns in our mind that dictate our life experiences. We can further heal trauma, gain a better understanding of ourselves, reset the body–mind connection, including our stress triggers, and stop the thought mechanisms that keep our breathing constricted.

But how? This is something Norm and I are still exploring.

We know that when you practice hyperventilation, one of the first things that happens is you alter your body's pH. You force carbon dioxide levels to drop and your blood becomes alkaline. When practiced deliberately and carefully, this reduction of carbon dioxide reduces the availability of oxygen to your tissues and cells, despite the fact that you're breathing more.

Some researchers believe this type of rapid breathing works as a form of exposure therapy, pushing the brain and body into a stressed panic state—but in a controlled environment and calm consciousness, this can cognitively teach the brain to deal with stress and panic without catastrophizing. This could be helpful for some. But, owing to my practice, I believe actually something else much more profound happens in these sessions.

After about 10-20 minutes of hyperventilation through conscious energy breathwork, we start to deactivate the brain's "default mode network."

The default mode network (DMN) is a group of brain structures found in the frontal and prefrontal cortex. It connects parts of the brain's thinking, decision-making and interpretative functions—which include the capacity for self-reflection, mental projection, past and future thought, and the ability to "interpret others"—with deeper and evolutionarily older structures of the brain that are involved in emotion and memory. The DMN is believed to be the locus of the ego and the home of rigid, habitual thinking and obsessions. It even lights up when given a list of adjectives related to a person's self-identity, and reacts in a similar way during self-reflection, memory recall and even when we get likes on social media. When there's no task at hand, the default mode network activates "by default" (hence its name) by doing things like daydreaming. Studies suggest that depression is linked to an overactive DMN. When your DMN is highly active, and you ruminate, you overanalyze yourself through the lens of your negative bias, and step out of the present moment to constantly question the past and the future.[80]

Accessing a less-than-ordinary state, whether it's brought about through under-stimulation, overstimulation breathing methods or by other means, has been shown to decrease blood flow and electrical activity in the DMN. When activity in the DMN falls off steeply, the ego temporarily vanishes, and the usual boundaries we experience between ourselves and the world around us, between subject and object, melt away.[81]

As I suggested earlier, once you've bypassed the ego, you're freed of your usual psychological defenses: the beliefs and patterns that, superficially, make you "you." The overthinking, obsessive part of your brain relaxes as you stop engaging in negative thought patterns, and you begin to experience a transcendence of space and time. In this state, you're free to explore your unconscious mind, and you connect to a world beyond your sensory perception. In effect, if your brain is like a house, then in this state you're handed the keys to the locked cellar. So any old repressed memories or experiences that are long forgotten can be accessed; there are no inhibiting barriers in place. This gives you an opportunity to process and integrate those memories and experiences, together with the energetic charge they possess within the body.

It has also been shown through EEG tests that the brain emits theta and delta waves during deliberate hyperventilation breathwork practice, suggesting the activation of creative and visionary abilities.[82] It also seems to be the case that through this type of breathwork, the stimulation of nerve cells is increased.

The important thing to understand about the ego is that the bigger it is, the more we think of ourselves as a distinct entity, separated from everything around us. The weaker our ego, the stronger our sense of connection with what's around us. We need the ego—it evolved to help us to stay alive. But we also need to keep it in check if we're to understand that, actually, we're closely connected with everything.

BEYOND TIME

Tick, tock. What happened yesterday? What's happening tomorrow? Humans are obsessed with time. Yet time is a construct of the mind. When we're stuck in time, in the past or future, we find it hard to function in the present. But what marks the passage of time in the mind?

Researchers at the Kavli Institute for Systems Neuroscience in Trondheim, Norway, have discovered a network of brain cells that expresses our sense of time within experiences and memories.[83] Although we know that the seconds on a clock always tick at the pace we've settled upon to measure time, clocks are devices created by humans. As social beings, we've decided to coordinate our activities according to an agreed-upon unit of time measurement. But your brain does not perceive time elapsing with the standardized units of hours and minutes. The signature of time in our experiences and memories is mapped differently in the brain.

If our breathing is mapped to our experiences and memories too, could breathing play a role in the brain's perception of time? It seems so. I've observed with hundreds of clients that when someone's breath is trapped, they fear taking the next breath because they don't want to move forward. Sometimes someone holds on to their breath because they don't want to let go—they're not trusting enough to let go of the past. They use their breathing to trap themselves in time.

In effect, you're handed the keys to the cellar.

WHY ARE THESE EXPERIENCES SO STRANGE?

Why did Ryan end up in a starlit cave? There are a few theories why certain spiritual experiences are so dreamlike and otherworldly. One fascinating reason for this could be linked to the same reason we're transported to another realm when we go to sleep, which some researchers suggest is due to a naturally occurring chemical called dimethyltryptamine (DMT).[84] DMT, an organic compound found in many animals and plants, is the main ingredient in many plant medicines used by a number of cultures for tens of thousands of years for healing, solving problems, and for connecting deeply with both their neighbors and the natural world.

Some experts believe this chemical is produced when we're in deep sleep, to create a sense of entering other realities. And just as dreams can be meaningful, often showing us areas of our life we need to think about more deeply or providing us with a particularly useful insight, some people, tribes and religious cultures have tried to induce a "waking dream" using DMT in a controlled environment. This is why DMT is often referred to as the "spirit molecule." Imperial College London has conducted research into this, as reported on the BBC.[85]

There's growing evidence that the lungs and the human brain have the capacity to make their own supply of DMT, and that it could potentially be released during breathwork sessions. Research in rats has shown that when they're highly stressed, their brains release high quantities of DMT. The same is true of human beings, who also release large amounts of DMT when close to death. So it could be that deliberately evoking the stress response in hyperventilation practice forces the body to produce it, although more research needs to be done to understand what levels of stress would be required to release DMT in quantities that could produce a spiritual experience.

Another link to the potential release of DMT during breathwork practice comes from research that shows high quantities of DMT helping brain cells survive in low-oxygen conditions. A study that involved giving a high dose of DMT to brain cells made them up to three times as likely to survive in oxygen levels of 0.5 percent, compared with our normal levels of 20 percent,[86] suggesting that we could potentially trigger this DMT release through hyperventilation, which causes carbon dioxide to fall and stops the release of oxygen from hemoglobin into the cell. Another route to low-oxygen conditions are practices that involve breath retention.

THE NEED FOR THE TRANSCENDENT

In simple language, transcendence is just going beyond our day-to-day experience and our usual concerns about ourselves, which are the products of an active ego.

The rising interest in breathwork, and in the potential of feelings of transcendence to help us with depression, PTSD and other psychological ailments that scar our lives, suggests an opening of the Western mind. During spiritual experiences, with our ego diminished, we have the ability, if only momentarily, to really rise above ourselves and think of ourselves as an integral part of our species, the planet and the universe, not just as it is now, but as it has always been and will be.

When I had my first experience of breathwork—the experience that changed me forever and led me down this path—I underwent a very powerful spiritual experience, a feeling of deep connectedness and transcendence, and it came with an intuition that Tiff was there beside me. If you asked me today whether I thought she was "really" present, I still wouldn't know. But what I'm beginning to find out, as so many others have found out too, is that whether she was or wasn't there doesn't really matter. I perceived that she was there, in real time, and the result was a transformative experience that sent me down a new, happier path.

EXERCISE 40

BIG VISION STATES FOR MANIFESTATION AND HEALING

At the end of your 10 minutes of Infinity Breathing with your humming sounds and affirmations, I want you to sit still for a few more minutes in a relaxed state. Reconnect to your intentions, your affirmations. Allow your breathing to come back to a natural slow rhythm and, in this theta state, let your imagination go. Start to think about the life you want to lead, the person you want to be, maybe the difference you want to make in the world. Perhaps there's a skill you've always wanted to master, maybe there's someone you've always wanted to meet, maybe there's a place you've always wanted to visit. Perhaps you want to be that person who uplifts everyone around you. Perhaps you've no clue for now—and that's OK.

- Find a comfortable spot, seated or lying down.
- Set a timer for 10 minutes, or have some music ready that will last about that long.
- Allow yourself time to anchor into your body.
- Notice your body, let it soften. Let go of any tension in your face, your jaw, your neck.
- Notice your mind, be aware of your thoughts, judgments and opinions.
- Feel your heartbeat in your chest and say your affirmation statements.

- Start your Infinity Breathing.
- Breathe in through your nose, feeling your belly rise.
- Breathe out through your nose. Relax and let go.
- Without a pause, breathe in. Open and expand.
- Breathe out. Relax and let go.
- Continue with this Infinity Breathing flow.
- Active inhalation, passive exhalation.
- Whenever you feel the desire to, drum your hands on your knees while making humming sounds. You should do this for three rounds before returning to the Infinity Breathing practice. Remember, if any emotion surfaces, allow yourself to feel it. If you feel uncomfortable, take a break.
- When the 10-minute timer is up, settle. Anchor yourself back into your body.
- Take some slow, deep, relaxed breaths.
- Feel your heartbeat again, maybe even place your hands on your heart.
- Feel the appreciation for your heart beating, for the life and vitality in your body.
- Feel gratitude for all the things that make you feel safe and loved—the people in your life, the resources and opportunities you have available to you.
- Come back to your heartbeat and now repeat your affirmations. Return to your affirmations, whatever they are: *I am strong, I am peaceful, I am proud, I am loving.*

- And don't just say them in your mind, really feel them in your body.
- Now, I want you to picture yourself as the ideal version of you.
- Transport yourself there.
- Where are you?
- What are you doing?
- What can you see?
- What are you wearing?
- How are you standing or sitting?
- What can you hear?
- What can you smell?
- What can you taste?
- Just let your mind wander.
- See if you can get glimpses of the ideal you.
- Can you feel the ground below you?
- Can you feel your heartbeat as this ideal you?
- Keep with it.
- Dream big!
- Who are you with?
- How do you feel?
- Keep with it if you can.
- Picture yourself.
- You're next level.

- You're amazing.
- Picture it right now in your mind.
- Feel it.
- Now I want you to be grateful for this vision.
- Feel the emotion of gratitude.
- You're 10 out of 10.
- How amazing and grateful you are.
- Really feel it.
- Take another deep breath in through your nose. Feeling your belly rise.
- And take a slow breath out through your mouth.
- One more. Deep breath in through your nose.
- And out through your mouth as you open your eyes.
- With as much detail as you can, write down and describe your big dream or vision.

Include the feelings this big vision creates in your body and mind.

You may want to use this as a new intention looking forward.

You can also seek experiences in your day that create more of those feelings. This will help to reprogram your body and mind to become open to the possibility of your big vision materializing.

★★★

A congratulations is in order. Over the last three chapters you have learned how you can use your breathing to thrive in all areas of your life, from work to sports. You've also learned how you can use your breathing to access transcendent states and feel more connected to the world around you. I said in the preface of this book that, through breathing, you could grow physically, mentally and emotionally. You now have all the knowledge and tools you need to ensure this happens. You understand the mechanics of your breathing—the importance of the nose and diaphragm, the function of the nervous system and the role of balance in the body. You know how your breathing can help you with difficult emotions, release the past and even deepen your connection with people around you. And you know how to optimize your breathing for different situations: for when you need to relax, focus, boost your energy, run your first marathon, or access visionary states.

To reap the reward of better breathing, you should continue your practice for 10 minutes per day. If you haven't already, you should make sure to complete your 40 days of Infinity Breathing. After your 40 days have elapsed, you may wish to continue using your Infinity Breathing practices to achieve a greater integration between your body, mind and emotions. Alternatively, you may wish to use your 10 minutes to explore your breath in any way you need; some days this might just involve practicing the magic ratio, while on others, you may use this 10 minutes to find your flow or perform better in sports. Now that you are acquainted with such a breadth of exercises, you should listen to your mind, feel your body and breath, and choose an exercise that will support your needs.

As we have explored throughout this book, breathing is a powerful and accessible tool to help you shift your state, calm your stress, reduce your pain or manage your simple emotions, whenever you need it. In particular, remember that "If in Doubt, Breathe It Out"! At difficult times in your life, the exercises in this book can give you strength and empower you to take control of your thoughts and feelings and let go of the grip of the past. And the more you come back to it, the more you immerse yourself in it, the better prepared you'll be for whatever life has in store for you in future.

If you want to take your breathwork practice forward even further, join us at www.breathpod.me or connect with me on my social channels @breathpod, where I bring together different approaches and types of breathwork delivered both online and in person to help people from every walk of life to flourish. There's just one thing left to say…

Although the book is almost over, your practice isn't.

TRAVEL LIGHT

When Tiff was diagnosed with cancer, she started keeping a blog to log her experiences along the journey.[87] It was a witty, honest, inspiring and often insightful record of somebody attempting to keep their spirits up in what were likely to be the final months of their life. One of the last things she ever wrote, she never posted. But it contained a grain of wisdom that I only remembered as I reached the end of writing this book: travel light.

Which, if you remember, was my intention at my first-ever breathwork session.

This is what this book is *really* about. It's about letting go of all your baggage so you can travel lightly through life. It's about casting off all your stress and anxiety, all your toxic beliefs and traumatic memories. It's about putting down that bag, taking control when you need to and letting go of that control, emptying your bag out, all at once, or brick by brick.

Life still happens. Sometimes you get some last-minute work on a tight deadline. Sometimes you just don't get a good night's sleep and that puts you in a bad mood. Sometimes you find that whatever you had for dinner doesn't really agree with you. But with the skills you now have, you can cruise through all that. And anyway, without the bad, you can't appreciate the good.

Of course, you'll also have experiences that are harder to deal with. As you move through life, you'll suffer emotional wounds. Things will throw you off balance. You might suffer loss and upheaval or feel deep grief. But now you know how to feel your way through that trauma, how to use your breath to integrate those difficult emotions. And what that means is that you can bounce back, move on, find solace. You don't have to restrict your feelings anymore. You don't have to let them take hold in your body. You don't have to freeze yourself in time just to feel safe. You can now work with your emotions. You can keep going forward.

Of course, your happiness doesn't just depend on what happens today or tomorrow. It also depends on what has come before. That's why we've explored how you can start to think about your past, and clean out those nicks, scrapes and deeper wounds so they can heal fully as well. As you now know, many of the problems we face in the present began in the past. That's why it's so important that we do that deeper work, to go beyond grappling with the stress and anxiety of the day.

We're living in a crazy but exciting time. We live in a world of smartphones, virtual reality, the metaverse, commercial space travel. We can order almost anything we want right to our door. And more people are taking care of themselves, physically and emotionally. We're more aware of our minds and bodies, our mental health and physical well-being. This is great, and breathing is one of an amazing number of tools to help us with this.

But there are problems too, and that's why it's more important than ever that we drop our bag of bricks—even that we start to empty it. There are global problems: we've neglected our planet and are now dealing with the consequences. And there are problems on our doorstep, like continued injustice and the mental health crisis. The truth is, in an ever-changing world like ours, there will always be challenges. And the problems we face as a species usually mirror our problems as individuals. If we don't take care of ourselves, we can't really take care of others and our planet. That's often how problems grow. So, like a parent who puts on their own oxygen mask before helping their child with theirs, we must remember that to help other people, we have to help ourselves first.

We all have it within ourselves to make a difference. Humans are capable of doing amazing things, as we demonstrate every single day in small ways and big. We perform a random act of kindness. We come up with a killer idea at work. We make a stranger laugh. We bounce back from adversity. And then there are the big ones: breaking world records, going into space, climbing the highest mountains on the planet. Humans are incredible.

Change starts with you. Right here, right now, it starts with you. Be the change you want to see in the world. You've the power to continue to make a difference, by cementing your daily breathing practice into your routine beyond this book so that you can continue to take control of your thoughts and emotions, rather than let them control you. By giving yourself permission to look inside yourself and address whatever you're going through and whatever you've been through over the course of your life, you can do something really special. It might not feel that way, but it's enough.

I want to end this book with a quote from one of my favorite philosophers, a personal hero and daily inspiration to me, Jiddu Krishnamurti:

What you are, the world is. And without your transformation, there can be no transformation of the world.

I hope you've enjoyed this book. Travel light.

★★★★★

ENDNOTES

1 BREATHING, THINKING AND FEELING

1. Joe Dispenza, *Breaking the Habit of Being Yourself: How to Lose Your Mind and Create a New One* (Hay House, 2012).

2 WHAT YOUR BREATHING SAYS ABOUT YOU

2. M. Thomas et al., "Prevalence of dysfunctional breathing in patients treated for asthma in primary care: cross sectional survey," *BMJ* (Clinical research ed.), 5 May 2001, vol. 322(7294), pp. 1098–100. doi:10.1136/bmj.322.7294.1098.

3. Dimitri Poddighe et al., "Non-allergic rhinitis in children: Epidemiological aspects, pathological features, diagnostic methodology and clinical management," *World Journal of Methodology*, 26 December 2016, vol. 6(4), pp. 200–13. doi:10.5662/wjm.v6.i4.200.

4. "The latest data on air quality and health where you live and around the globe," https://www.stateofglobalair.org.

5. "Climate-related variation of the human nasal cavity," https://pubmed.ncbi.nlm.nih.gov/21660932/.

6. H. J. Schünemann et al., "Pulmonary function is a long-term predictor of mortality in the general population: 29-year

follow-up of the Buffalo Health Study," *Chest*, September 2000, vol. 118(3), pp. 656–64. doi: 10.1378/chest.118.3.656.

7. Brian K. Rundle et al., "Contagious yawning and psychopathy," *Personality and Individual Differences*, November 2015, vol. 86, pp. 33–7. doi.org/10.1016/j.paid.2015.05.025.

3 SHUT YOUR MOUTH AND SLOW YOUR FLOW

8. Dr. Alan Ruth, "The Health Benefits of Nose Breathing," https://www.lenus.ie/bitstream/handle/10147/559021/JAN15Art7.pdf;jsessionid=FF4B506CE0A8763DC8EA9F68D86C686C?sequence=1.

9. S. Naftali et al., "The air-conditioning capacity of the human nose," *Annals of Biomedical Engineering*, April 2005, vol. 33, pp. 545–53. doi: 10.1007/s10439-005-2513-4.

10. Robin L. Rothenberg, *Restoring Prana: A Therapeutic Guide to Pranayama and Healing Through the Breath for Yoga Therapists, Yoga Teachers and Healthcare Practitioners* (Singing Dragon, 2019), p. 115.

11. M. H. Cottle, "Nasal breathing pressures and cardio-pulmonary illness," *Eye, Ear, Nose and Throat Monthly*, September 1972, vol. 51(9), pp. 331-340, https://pubmed.ncbi.nlm.nih.gov/5068888/.

12. "The Nobel Prize in Physiology or Medicine 1998," Nobel Prize Outreach AB 2022, Friday 4 March 2022. NobelPrize.org.

13. Alexi Cohan, "Nitric oxide, a 'miracle molecule,' could treat or even prevent coronavirus, top doctors say," https://www.bostonherald.com/2020/07/26/nitric-oxide-a-miracle-molecule-could-treat-or-even-prevent-coronavirus-top-doctors-say/.

14. P. Barnes, "NO or no NO in asthma?" https://pubmed.ncbi.nlm.nih.gov/8711662/.

15. Uppsala University, "Nitric oxide a possible treatment for COVID-19, study finds," *ScienceDaily*, 3 October 2020.

16. Dario Akaberi et al., "Mitigation of the replication of SARS-CoV-2 by nitric oxide in vitro," *Redox Biology*, 2020, vol. 37, p. 101734. doi: 10.1016/j.redox.2020.101734.

17. K. Upadhyay-Dhungel and A. Sohal, "Physiology of nostril breathing exercises and its probable relation with nostril and

cerebral dominance: A theoretical research on literature," *Janaki Medical College Journal of Medical Science*, 2003, vol. 1(1), pp. 38–47. doi: 10.3126/jmcjms.v1i1.7885.

18. G. F. Karliczek et al., "Vasoconstriction following neuroleptanesthesia. Hemodynamic studies after open heart surgery," *Acta Anaesthesiologica Belgica*, 1979, vol. 30, pp. 213–31.

19. Yogananda, *Autobiography of a Yogi*, p. 240, 2006 edition.

20. Marc A. Russo et al., "The physiological effects of slow breathing in the healthy human," *Breathe*, December 2017, vol. 13(4), pp. 298–309. doi: 10.1183/20734735.009817.

21. I. M. Lin et al., "Breathing at a rate of 5.5 breaths per minute with equal inhalation-to-exhalation ratio increases heart rate variability," *International Journal of Psychophysiology*, March 2014, vol. 91(3), pp. 206–11. doi: 10.1016/j.ijpsycho.2013.12.006.

22. Bernardi et al., "Effect of rosary prayer and yoga mantras on autonomic cardiovascular rhythms: comparative study," https://pubmed.ncbi.nlm.nih.gov/11751348/.

23. V. Müller and U. Lindenberger, "Cardiac and respiratory patterns synchronize between persons during choir singing," *PLOS ONE*, September 2011, vol. 6(9). doi: 10.1371/journal.pone.0024893.

24. https://buteyko.ru/eng/interw.shtml.

4 STRESS LESS, SLEEP BETTER AND MANAGE PAIN

25. Mental Health Foundation (2018), "Stressed nation: 74% of UK 'overwhelmed or unable to cope' at some point in the past year," www.mentalhealth.org.uk/news/stressed-nation-74-uk-overwhelmed-or-unable-cope-some-point-past-year.

26. Maureen Connolly and Margot Slade, "The United States of Stress 2019," *Everyday Health*, October 2018, www.everydayhealth.com/wellness/united-states-of-stress/.

27 Questions 1 and 2 taken from Tim Ferris, "5 Morning Rituals that Help Me Win the Day" (July 2018). https://tim.blog/wp-content/uploads/2018/07/5-morning-rituals-that-help-me-win-the-day-july2018.pdf.

28. S. W. Porges, "Orienting in a defensive world: Mammalian modifications of our evolutionary heritage. A Polyvagal theory," *Psychophysiology*, 32:4 (July 1995), pp. 301–18. doi: 10.1111/j.1469-8986.1995.tb01213.x.

29. Swapna Bhaskar et al., "Prevalence of chronic insomnia in adult patients and its correlation with medical comorbidities," https://www.ncbi.nlm.nih.gov/pmc/articles/PMC5353813/.

30. S. S. Campbell et al., "Alleviation of sleep maintenance insomnia with timed exposure to bright light," *Journal of the American Geriatrics Society*, August 1993, vol. 41(8), pp. 829–36. doi: 10.1111/j.1532-5415.1993.tb06179.x.

31. https://hubermanlab.com/dr-matthew-walker-the-science-and-practice-of-perfecting-your-sleep/.

32. https://www.medicaldaily.com/life-hack-sleep-4-7-8-breathing-exercise-will-supposedly-put-you-sleep-just-60-332122.

33. https://www.ncbi.nlm.nih.gov/pmc/articles/PMC7007763/.

34. V. Busch et al., "The effect of deep and slow breathing on pain perception, autonomic activity, and mood processing—an experimental study," *Pain Medicine*, February 2012, vol. 13(2), pp. 215–28. doi: 10.1111/j.1526-4637.2011.01243.x.

35. Harvard Health Publishing, Harvard Medical School, "Six ways to use your mind to control pain," April 2015, www.health.harvard.edu/mind-and-mood/6-ways-to-use-your-mind-to-control-pain.

5 UNDERSTAND YOUR EMOTIONS

36. Juan Murube, "Hypotheses on the development of psychoemotional tearing," *The Ocular Surface*, October 2009, vol. 7(4), 2009, pp. 171–5. doi: 10.1016/S1542-0124(12)70184-2.

37. https://www.ncbi.nlm.nih.gov/pmc/articles/PMC4035568/.

38. Anja J. Laan et al., "Individual differences in adult crying: the role of attachment styles," *Social Behavior and Personality*, April 2012, vol. 40(3), pp. 453–71. doi: 10.2224/sbp.2012.40.3.453.

39. Candace Pert, *Molecules of Emotion: Why You Feel the Way You Feel* (Simon & Schuster, 1999).

40. https://pubmed.ncbi.nlm.nih.gov/26893293/.

41. https://www.ncbi.nlm.nih.gov/pmc/articles/PMC5546756/.

42. Lauri Nummenmaa et al., "Bodily maps of emotions," *Proceedings of the National Academy of Sciences*, January 2014, vol. 111(2), pp. 646–51. doi: 10.1073/pnas.1321664111.

43. Candace Pert, *Everything You Need to Know to Feel Go(o)d* (Hay House, 2007).

6 RELEASE TRAUMA AND REWIRE YOUR MIND

44. https://www.youtube.com/watch?v=nmJOuTAk09g.

45. If you are interested in learning more about Orr's work, I would recommend that you read *Rebirthing in the New Age* (Trafford Publishing, 2007) or *Manual for Rebirthers: How to deepen your Rebirthing process, Masterfully guide other people's process and be a successful Rebirthing Professional* (Vision Libros, 2011). You can also visit https://www.rebirthingbreathwork.com/.

8 FIND FLOW, FOCUS AND ENERGY

46. Mihaly Csikszentmihalyi, *Flow: The Psychology of Happiness* (Rider Books, 1990).

47. Susie Cranston and Scott Keller, "Increasing the meaning quotient of work," *McKinsey Quarterly* 1 (2013), pp. 48-59.

48. https://web.archive.org/web/20150206014318/; https://www.cc.gatech.edu/~vector/papers/sqj.pdf.

49. https://www.forbes.com/sites/johnhall/2020/05/03/the-biggest-culprit-behind-your-lagging-productivity-you/?sh=40a54ee97625.

50. *From Destructive Emotions: How Can We Overcome Them?* Narrated by Daniel Goleman. © 2003 by Mind and Life Institute. Published by arrangement with Bantam Books.

51. http://citeseerx.ist.psu.edu/viewdoc/download?doi=10.1.1.867.3694&rep=rep1&type=pdf; https://www.researchgate.net/publication/16581710_Stimulus_control_applications_to_the_treatment_of_worry.

52. https://www.ncbi.nlm.nih.gov/pmc/articles/PMC6852150/.

53. https://www.health.harvard.edu/mind-and-mood/protect-your-brain-from-stress#:~:text=%22A%20life%20without%20stress%20is,for%20healthier%20responses%20to%20stress.

54. Graham Wallas, *The Art of Thought* (Harcourt, Brace and Co., 1928).

55. https://philadelphia.cbslocal.com/2013/04/12/study-80-of-people-grab-smartphone-within-15-minutes-of-waking/#:~:text=Study%3A%2080%25%20Of%20People%20Grab%20Smartphone%20Within%2015%20Minutes%20Of%20Waking,-April%2012%2C%202013&text=PHILADELPHIA%20(CBS)%20%E2%80%93%20A%20new,own%20iPhones%20or%20Android%20smartphones.

56 Scott Kaufman and Carolyn Gregoire, *Wired to Create: Unravelling the Mysteries of the Creative Mind* (Vermilion, 2016).

57. https://www.dovepress.com/assessing-public-speaking-fear-with-the-short-form-of-the-personal-rep-peer-reviewed-fulltext-article-NDT.

58. https://www.mkgandhi.org/own_wrds/own_wrds.htm.

9 IMPROVE YOUR FITNESS

59 Matt Slater, "Olympics cycling: Marginal gains underpin Team GB dominance" (8 August 2012). BBC Sport.

60. https://www.youtube.com/watch?v=MGkuxFY0lyQ.

61. George Dallam et al., "Effect of nasal versus oral breathing on VO_2 max and physiological economy in recreational runners following an extended period spent using nasally restricted breathing," *International Journal of Kinesiology and Sports Science*, April 2018, vol. 6(22), pp. 22–9. doi: 10.7575/aiac.ijkss.v.6n.2p.22.

62. https://thorax.bmj.com/content/60/8/629.

63. https://www.ncbi.nlm.nih.gov/pmc/articles/PMC4267026/.

64. Patrick McKeown, *The Oxygen Advantage: The Simple, Scientifically Proven Breathing Techniques for a Healthier, Slimmer, Faster, and Fitter You* (Piatkus, 2015).

65. https://pubmed.ncbi.nlm.nih.gov/677559/.

66. "Profiles: Emil Zátopek," Running Past. http://www.runningpast.com/emil_zatopek.htm.

67. Pascal Mollard et al., "Validity of arterialized earlobe blood gases at rest and exercise in normoxia and hypoxia," *Respir Physiol Neurobiol*, 172:179-183, 2010.

68. https://pubmed.ncbi.nlm.nih.gov/29400616/.

69. https://www.ncbi.nlm.nih.gov/pmc/articles/PMC2771546/.

70. https://www.ncbi.nlm.nih.gov/pmc/articles/PMC3361916/.

71. https://www.ncbi.nlm.nih.gov/pmc/articles/PMC3756401/.

72. https://training.seer.cancer.gov/disease/categories/classification.html#:~:text=Carcinomas%2C%20malignancies%20of%20epithelial%20tissue,such%20as%20the%20gastrointestinal%20tract.

73. Mary Mohrin et al., "Stem cell aging: A mitochondrial UPR-mediated metabolic checkpoint regulates hematopoietic stem cell aging," *Science*, March 2015, vol. 347(6228), pp. 1374–7. doi: 10.1126/science.aaa2361.

74. https://www.ncbi.nlm.nih.gov/books/NBK536728/.

75. Katharina Lust and Joachim Wittbrodt, "Hold your breath!", *eLife*, December 2015, vol. 4, e12523. doi: 10.7554/eLife.12523.

76. Matthijs Kox et al., "Voluntary activation of the sympathetic nervous system and attenuation of the innate immune response in humans," *Proceedings of the National Academy of Sciences of the United States of America*, May 2014, vol. 111(20), pp. 7379–84. doi: 10.1073/pnas.1322174111.

77. https://pubmed.ncbi.nlm.nih.gov/22685240/; https://clinicaltrials.gov/ct2/show/NCT01835457; https://pubmed.ncbi.nlm.nih.gov/24799686/.

78. E. Schagatay et al., "Underwater working times in two groups of traditional apnea divers in Asia: the Ama and the Bajau," *Diving and Hyperbaric Medicine*, March 2011, vol. 41(1), pp. 27–30. PMID: 21560982.

10 ACCESS TRANSCENDENT STATES

79. If you are interested in further exploring Grof's work, you can find a list of books on his website: https://www.stangrof.com/index.php/books. In particular, I would recommend reading: *Holotropic Breathwork: A New Approach to Self-Exploration and Therapy* (Excelsior Editions, 2010), *Psychology of the Future: Lessons from Modern Consciousness Research* (State University of New York Press, 2000) and *The Cosmic Game: Explorations of the Frontiers of Human Consciousness* (State University of New York Press, 1997). Additionally, *The Tim Ferriss Show* podcast, #347: "Stan Grof, Lessons from ~4,500 LSD Sessions and Beyond," is also excellent.

80. https://www.nature.com/articles/srep43105.

81. M. Pollan, *How to Change Your Mind: What the New Science of Psychedelics Teaches Us About Consciousness, Dying, Addiction, Depression, and Transcendence* (Penguin Books, 2019).

82. https://www.ncbi.nlm.nih.gov/pmc/articles/PMC3952317/; https://www.oatext.com/eeg-response-to-hyperventilation-in-patients-with-cns-disorder.php.

83. Albert Tsao et al., "Integrating time from experience in the lateral entorhinal cortex," *Nature* (30 August 2018). Kavli Institute for Systems Neuroscience and Centre for Neural Computation, NTNU, Trondheim, Norway.

84. https://www.nature.com/articles/s41598-019-51974-4?fbclid=IwAR1QOx0Dwzm8XXLYhJAj6vI4_SD_zfULw6SDUfFAQY0OuzNj-ma13ZAokDc.

85. https://www.bbc.co.uk/news/health-56373202.

86. https://www.ncbi.nlm.nih.gov/pmc/articles/PMC5021697/.

TRAVEL LIGHT

87. Tiff's blog can still be found at: https://ittybittycancertittycommittee.tumblr.com/.

INDEX

Exercise names and page numbers for illustrations are *italicized*.

ACKNOWLEDGMENTS

I never ever thought I would or could write a book, so this has been a journey of introspection and a real roller coaster for me. This book certainly wouldn't be in your hand without a whole tribe of people. Thanks to all the breathwork schools, teachers and facilitators who have helped and inspired me greatly, especially Anne-Marie, Pippa & Russell at Edinburgh Breathes, Judith Kravitz, Eugenia Altamira and all the trainers at Transformational Breath Foundation and of course all the "Respiremos" family in Mexico. Thanks to Stan Grof, Leonard Orr and Kriya Yoga masters. Thanks to Patrick McKeown, Wim Hof (and daily cold showers) and everyone at The International Breathwork Foundation. Thank you to all my other teachers, mentors, coaches, breathworkers, instructors, doctors, researchers, gurus, yogis, guides, authors, podcasters, friends and cheerleaders who have supported me along the way. Some were intricately involved, and some were involved but don't even know it. Even you, now, with the book in your hand, played your part—so thank you.

A deeply felt thanks to every client who has walked through my door and everyone who has joined me at events both online or in person. I have learned, and continue to learn, so much from you all. It has been my pleasure to help you to overcome your challenges and move forward in your lives.

Extra-special shout-outs go to:

My dearest partner Nova, who has accompanied me on this journey. I am deeply grateful for your daily support, companionship and for showing me what love is again. Thank you for your patience and for putting up with "Book Stu," which at times has been a whole other stratosphere of stress beyond "Airport Stu."

Harry Readhead—man, I don't know what I would have done without you. Your energy is felt on these pages. Thanks for making me think differently, see the world from new perspectives and inspiring me to write. You have helped me beyond belief through this process and kept me breathing calmly throughout.

Deep thanks to Bev James. Thank you for always being there for me, believing in me and not taking my "No, I can't write a book!" as an answer. And a special thanks to the brilliant BJM team too: Tom Wright, Morwenna Loughman, Serena Murphy and all the gang, for your input, support and guidance. Also, Amy Warren, who helped me to find my feet with writing at the start.

Thank you to the team of pros at HQ and HarperCollins—editors Abigail Le Marquand-Brown, Laura Bayliss, Kate Fox and Mark Bolland for shaping this book into its current form. Authors don't get to see all the cogs in the machine that contributes to getting a book on the shelves, but I am so grateful to everyone who has done their part behind the scenes. Thanks for trusting me and believing in my vision.

Huge thanks to Belle PR for your love and support, and Luxley PR for believing in me, and helping us spread our message far and wide.

So much love to Harry Pearce and Protein Studios for bringing the good vibes every day, and lending me their Harry Potter cupboard to write in.

Thanks my dear friend Dr. Norm for your brilliant mind and infectious laugh, that always lights up the room. It was your encouragement that made me look deeper into these incredible teachings through my own research. I feel extremely lucky to have your support.

Big thanks to David Johnston and Tom Sharp for their creative minds, inspiring leadership and the whole team at Accept and Proceed.

And finally, a huge heartfelt thanks to my parents, Neil and Joyce Sandeman. You have always believed in me, supported me and encouraged me to believe in myself. I love you and the ketchup on the wall.